T0313834

MITIGATING TIN WHISKER RISKS

WILEY SERIES ON PROCESSING OF ENGINEERING MATERIALS

Randall M. German, Editor

Structure - Property Relations in Nonferrous Metals
Alan M. Russell and Kok Loong Lee

Handbook of Mathematical Relations In Particulate Materials Processing
Randall M. German and Seong Jin Park

Chemistry and Physics Of Mechanical Hardness
John J. Gilman

Principles of Laser Materials Processing
Elijah Kannatey - Asibu, Jr.

Lecture Topics in Materials Thermodynamics
Y. Austin Chang and W. Alan Oates

Materials Thermodynamics
Y. Austin Chang and W. Alan Oates

MITIGATING TIN WHISKER RISKS

Theory and Practice

TAKAHIKO KATO
CAROL A. HANDWERKER
JASBIR BATH

Copyright © 2016 by John Wiley & Sons, Inc. All rights reserved

Published by John Wiley & Sons, Inc., Hoboken, New Jersey
Published simultaneously in Canada

No part of this publication may be reproduced, stored in a retrieval system, or transmitted in any form or
by any means, electronic, mechanical, photocopying, recording, scanning, or otherwise, except as
permitted under Section 107 or 108 of the 1976 United States Copyright Act, without either the prior
written permission of the Publisher, or authorization through payment of the appropriate per-copy fee to
the Copyright Clearance Center, Inc., 222 Rosewood Drive, Danvers, MA 01923, (978) 750-8400, fax
(978) 750-4470, or on the web at www.copyright.com. Requests to the Publisher for permission should
be addressed to the Permissions Department, John Wiley & Sons, Inc., 111 River Street, Hoboken, NJ
07030, (201) 748-6011, fax (201) 748-6008, or online at http://www.wiley.com/go/permissions.

Limit of Liability/Disclaimer of Warranty: While the publisher and author have used their best efforts in
preparing this book, they make no representations or warranties with respect to the accuracy or
completeness of the contents of this book and specifically disclaim any implied warranties of
merchantability or fitness for a particular purpose. No warranty may be created or extended by sales
representatives or written sales materials. The advice and strategies contained herein may not be suitable
for your situation. You should consult with a professional where appropriate. Neither the publisher nor
author shall be liable for any loss of profit or any other commercial damages, including but not limited to
special, incidental, consequential, or other damages.

For general information on our other products and services or for technical support, please contact our
Customer Care Department within the United States at (800) 762-2974, outside the United States at
(317) 572-3993 or fax (317) 572-4002.

Wiley also publishes its books in a variety of electronic formats. Some content that appears in print may
not be available in electronic formats. For more information about Wiley products, visit our web site at
www.wiley.com.

Library of Congress Cataloging-in-Publication Data has been applied for

Hardback: 9780470907238

Typeset in 10/12pt TimesLTStd by SPi Global, Chennai, India

Printed in the United States of America

10 9 8 7 6 5 4 3 2 1

CONTENTS

List of Contributors   ix

Introduction   xi

1 A Predictive Model for Whisker Formation Based on Local
   Microstructure and Grain Boundary Properties   1
   Pylin Sarobol, Ying Wang, Wei-Hsun Chen, Aaron E. Pedigo, John P. Koppes,
   John E. Blendell and Carol A. Handwerker

   1.1 Introduction, 1
   1.2 Characteristics of Whisker and Hillock Growth from Surface Grains, 3
   1.3 Summary and Recommendations, 17
        Acknowledgments, 18
        References, 19

2 Major Driving Forces and Growth Mechanisms for Tin Whiskers   21
   Eric Chason and Nitin Jadhav

   2.1 Introduction, 21
   2.2 Understanding the Mechanisms Behind Imc-Induced Stress Evolution and
        Whisker Growth, 24
   2.3 Relation of Stress to Whisker Growth, 34
   2.4 Conclusions, 39
        Acknowledgments, 40
        References, 40

3 Approaches of Modeling and Simulation of Stresses in Sn Finishes   **43**
Peng Su and Min Ding

3.1 Introduction, 43
3.2 Constitutive Model, 44
3.3 Strain Energy Density, 46
3.4 Grain Orientation, 46
3.5 Finite Element Modeling of Triple-Grain Junction, 48
3.6 Finite Element Modeling of Sn Finish with Multiple Grains, 55
References, 66

4 Properties and Whisker Formation Behavior of Tin-Based Alloy Finishes   **69**
Takahiko Kato and Asao Nishimura

4.1 Introduction, 69
4.2 General Properties of Tin-based Alloy Finishes (Asao Nishimura), 70
4.3 Effect of Alloying Elements on Whisker Formation and Mitigation (Asao Nishimura), 75
4.4 Dependence of Whisker Propensity of Matte Tin–Copper Finish on Copper Lead-Frame Material (Takahiko Kato), 89
4.5 Conclusions, 118
Acknowledgments, 118
References, 119

5 Characterization Techniques for Film Characteristics   **125**
Takahiko Kato and Yukiko Mizuguchi

5.1 Introduction, 125
5.2 TEM (Takahiko Kato), 125
5.3 SEM (Yukiko Mizuguchi), 140
5.4 EBSD (Yukiko Mizuguchi), 146
5.5 Conclusions, 154
Acknowledgments, 155
References, 155

6 Overview of Whisker-Mitigation Strategies for High-Reliability Electronic Systems   **159**
David Pinsky

6.1 Overview of Tin Whisker Risk Management, 159
6.2 Details of Tin Whisker Mitigation, 164
6.3 Managing Tin Whisker Risks at the System Level, 173
6.4 Control of Subcontractors and Suppliers, 183
6.5 Conclusions, 185
References, 185

7   **Quantitative Assessment of Stress Relaxation in Tin Films by the
    Formation of Whiskers, Hillocks, and Other Surface Defects**   187
    *Nicholas G. Clore, Dennis D. Fritz, Wei-Hsun Chen, Maureen E. Williams,
    John E. Blendell and Carol A. Handwerker*

    7.1 Introduction, 187
    7.2 Surface-Defect Classification and Measurement Method, 189
    7.3 Preparation and Storage Conditions of Electroplated Films on
         Substrates, 194
    7.4 Surface Defect Formation as a Function of Tin Film Type, Substrate, and
         Storage Condition, 195
    7.5 Conclusions, 209
         Appendix, 209
         Acknowledgments, 209
         References, 213

8   **Board Reflow Processes and their Effect on Tin Whisker Growth**   215
    Jasbir Bath

    8.1 Introduction, 215
    8.2 The Effect of Reflowed Components on Tin Whisker Growth in Terms of
         Grain Size and Grain Orientation Distribution, 215
    8.3 Reflow Profiles and the Effect on Tin Whisker Growth, 216
    8.4 Influence of Reflow Atmosphere and Flux on Tin Whisker Growth, 219
    8.5 Effect of Solder Paste Volume on Component Tin Whisker Growth during
         Electronics Assembly, 220
    8.6 Conclusions, 221
         Acknowledgments, 222
         References, 222

9   **Mechanically Induced Tin Whiskers**   225
    Tadahiro Shibutani and Michael Osterman

    9.1 Introduction, 225
    9.2 Overview of Mechanically Induced Tin Whisker Formation, 227
    9.3 Theory, 228
    9.4 Case Studies, 237
    9.5 Conclusions, 245
         References, 246

Index   249

LIST OF CONTRIBUTORS

Jasbir Bath; Bath Consultancy LLC, San Ramon, California, USA

John E. Blendell; School of Materials Engineering, Purdue University, West Lafayette, Indiana, USA

Eric Chason; Divison of Engineering, Brown University, Providence, Rhode Island, USA

Wei-Hsun Chen; Cymer, Inc., San Diego, California, USA; School of Materials Engineering, Purdue University, West Lafayette, Indiana, USA

Nicholas G. Clore; School of Materials Engineering, Purdue University, West Lafayette, Indiana, USA

Min Ding; Freescale Semiconductor, Austin, Texas, USA

Dennis D. Fritz; Science Applications International Corporation (SAIC), McLean, Virginia, USA

Carol A. Handwerker; School of Materials Engineering, Purdue University, West Lafayette, Indiana, USA

Nitin Jadhav; IBM, Hopewell Junction, New York, USA

Takahiko Kato; Research & Development Group, Hitachi, Ltd., Tokyo, Japan; Center for Advanced Research of Energy and Materials, Hokkaido University, Sapporo, Japan

John P. Koppes; Alcoa-Howmet, Whitehall, Michigan, USA

Yukiko Mizuguchi; Advanced Materials Laboratory, Sony Corporation, Atsugi, Kanagawa, Japan

Asao Nishimura; Jisso Partners, Inc., Tokyo, Japan

Michael Osterman; Center of Advanced Life Cycle Engineering (CALCE), University of Maryland, College Park, Maryland, USA

Aaron E. Pedigo; Naval Surface Warfare Center, Crane Division, Crane, Indiana, USA

David Pinsky; Raytheon Integrated Defense Systems, Tewksbury, Massachusetts, USA

Pylin Sarobol; Sandia National Laboratories, Albuquerque, New Mexico, USA

Tadahiro Shibutani; Yokohama National University, Yokohama, Kanagawa, Japan

Peng Su; Juniper Networks, Sunnyvale, California, USA

Ying Wang; School of Materials Engineering, Purdue University, West Lafayette, Indiana, USA

Maureen E. Williams; National Institute of Standards and Technology (NIST), Gaithersburg, Maryland, USA

INTRODUCTION

The past few years have seen major research and development in the study of and understanding of tin whiskers with the continuing transition to lead-free assembly. This book covers key tin whisker topics, ranging from fundamental science to practical mitigation strategies, written by international experts in these areas. The chapters listed here provide concise analyses of the current state of the art in the following tin whisker subjects:

A Predictive Model for Whisker Formation Based on Local Microstructure and Grain Boundary Properties

Major Driving Forces and Growth Mechanisms for Tin Whiskers

Approaches of Modeling and Simulation of Stresses in Tin Finishes

Properties and Whisker Formation Behavior of Tin-Based Alloy Finishes

Characterization Techniques for Film Characteristics

Overview of Whisker Mitigation Strategies for High-Reliability Electronic Systems

Quantitative Assessment of Stress Relaxation in Tin Films by the Formation of Whiskers, Hillocks, and Other Surface Defects

Board Reflow Processes and Their Effect on Tin Whisker Growth

Mechanically Induced Tin Whiskers
    Chapter 1 reviews the characteristic properties of local microstructures around whisker and hillock grains with the aim to further identify why these particular grains and locations become predisposed to forming whiskers and hillocks.

On the local level, factors including surface grain geometry, crystallographic orientation-dependent surface grain boundary structure, and the localization of elastic strain/strain energy density distribution are discussed. A growth model is also discussed with a review on how it may be possible to engineer a whisker-resistant texture for the appropriate aging condition.

The focus of Chapter 2 is on describing the primary driving forces and mechanisms causing the growth of whiskers and hillocks. It describes the results of experiments that have been designed to distinguish among different potential mechanisms and identify the most important ones so as to develop a fuller understanding of the materials science behind whisker growth. It reviews how the growth of intermetallic compounds (IMC) leads to stress in the tin layer. Results from real-time studies of whisker growth in the scanning electron microscope/focused ion-beam milling (SEM/FIB) are also shown that review how whiskers and hillocks evolve in time.

Chapter 3 discusses the approach of finite element modeling for studying stress in a tin finish, which would lead to tin whiskers. Examples are given to explore the impact of the grain orientation on whisker growth in tin finishes.

Chapter 4 reviews the basic properties of tin-based alloy finishes and the effects of various alloying elements on whisker formation. The focus is on whisker test data and potential mechanisms for whisker suppression or enhancement for each element. Review of experimental data on the mechanisms of spontaneous whisker formation or suppression in matte tin–copper alloy finishes reveals how adding minor elements to the copper base material (lead frame) can significantly change the whisker formation propensity of the alloy finish.

Chapter 5 discusses characterization methods that can be used to easily clarify the essential characteristics of tin and tin-based alloy finishes and thus deepen our understanding of whisker formation mechanisms. These methods include transmission electron microscopy (TEM), scanning electron microscopy (SEM), and electron backscatter diffraction (EBSD).

Chapter 6 includes tin whisker risk mitigation strategies for each tier of the supply chain for high-reliability electronic systems and how each tier of the supply chain including metal finisher/component packagers, component packaging designer, assembly designer, subsystem assembler, systems designer and integrator, end user needs to play its proper role. The chapter also discusses whisker risk management techniques including system algorithm tools as a means to define semiquantitative design rules.

Chapter 7 focuses on a study to develop and validate a surface-defect counting procedure to be applied in research on the specific mechanisms responsible for stress relaxation and tin whisker formation in tin films. The methodology developed assessed differences in defect morphologies, dimensions, densities, and the total volume relaxed by particular types of defects including a comparison of the total defect volumes and the corresponding volumetric strains relaxed by defect formation. The methodology provided a means to quantify the complexity of stress relaxation in tin surface finishes.

Chapter 8 reviews the effects of reflow profile, reflow atmosphere, solder paste volume, and flux activity on tin whisker growth.

Chapter 9 discusses pressure-induced whiskers at separable interfaces for lead-free separable contacts and connectors. Theories of mechanically induced tin whiskers are reviewed with case studies using pure tin and lead-free finishes shown to evaluate the pressure-induced tin whiskers.

We hope that this book will provide members of the broader electronic packaging community with useful information and insights into whisker risk and mitigation. Armed with the insights presented here by scientists and engineers who have both breadth and depth of experience in investigating tin whisker formation, readers will be able to understand key factors and develop rational strategies for mitigation based on their own situations.

1

A PREDICTIVE MODEL FOR WHISKER FORMATION BASED ON LOCAL MICROSTRUCTURE AND GRAIN BOUNDARY PROPERTIES

PYLIN SAROBOL

Sandia National Laboratories, Albuquerque, New Mexico, USA

YING WANG

School of Materials Engineering, Purdue University, West Lafayette, Indiana, USA

WEI-HSUN CHEN

School of Materials Engineering, Purdue University, West Lafayette, Indiana, USA; Cymer, Inc., San Diego, California, USA

AARON E. PEDIGO

Naval Surface Warfare Center, Crane Division, Crane, Indiana, USA

JOHN P. KOPPES

Alcoa-Howmet, Whitehall, Michigan, USA

JOHN E. BLENDELL AND CAROL A. HANDWERKER

School of Materials Engineering, Purdue University, West Lafayette, Indiana, USA

1.1 INTRODUCTION

For over 60 years, whiskers and hillocks have been recognized as local manifestations of stress relaxation in thin films: these "surface defects" grow spontaneously by mass transport to specific grain boundaries in the plane of the film, with whiskers becoming

Mitigating Tin Whisker Risks: Theory and Practice, First Edition.
Edited by Takahiko Kato, Carol A. Handwerker, and Jasbir Bath.
© 2016 John Wiley & Sons, Inc. Published 2016 by John Wiley & Sons, Inc.
Companion website: www.wiley.com/go/Kato/TinWhiskerRisks

hillocks when grain boundary migration accompanies growth out of the plane of the film. Whisker formation is important for two reasons.

First, the risk to electronic system reliability is serious for whisker formation in Sn films: whiskers can sometimes grow to be millimeters long, causing short-circuiting between adjacent components. Second, isolating the mechanisms leading to whisker and hillock formation in one grain out of 10^3–10^5 film grains will inform us more broadly about the multiplicity of possible responses of thin films to stresses.

As a local phenomenon, whisker formation has been attributed in the past to specific inhomogeneity in either the film or the interfacial intermetallic microstructure, in stress, or in the anisotropic properties in the film. These inhomogeneities include

- grains above large Cu_6Sn_5 intermetallic particles [1, 2];
- grains covered by relatively weak oxide films [3];
- grains with larger out-of-plane elastic deformation than surrounding grains due to elastic anisotropy under plane strain conditions [4];
- a difference in the orientation-dependent yielding behavior of whisker grains and their neighboring grains [5, 6];
- shallow grains whose grain boundaries serve as sinks for atoms when surface diffusion necessary for diffusional creep is not active [7].

When the first three of these concepts were tested experimentally, these simple relationships were not observed. For example, Pei et al. [8] determined from electron backscatter diffraction (EBSD) and etching studies that whiskers were not associated with large underlying intermetallic particles. Moon et al. [9], Jadhav et al. [10], and A.E. Pedigo (unpublished research) determined through ultra-high vacuum (UHV) studies of sputtered films and through oxide removal in specific regions by focused ion beam (FIB) milling that whisker growth was unchanged when the native oxide layer was removed through sputtering and was not accelerated in regions without oxide.

From these and other experiments on the crystallography, texture, geometry, and strain in whisker and hillock grains relative to their surrounding microstructures, it is clear that models ascribing whisker formation and growth to a single cause are too simplistic: they do not capture all relevant aspects of why that particular grain was predisposed to become a whisker out of 10^3–10^5 grains in the film nor do they provide quantitative predictions of growth rates under a wide range of conditions.

Whisker nucleation and growth must be examined at the local level, taking into consideration the microstructure and properties of the film, specifically grain boundary structure, as well as crystallography, grain geometry, and the role of oxide films.

In this chapter, we review the characteristic properties of the local microstructure around whisker and hillock grains with the aim to further identify why these particular grains and locations became predisposed to forming whiskers and hillocks.

On the local level, we suggest that three factors play major roles in determining which surface grains will become whiskers or hillocks: the surface grain geometry, crystallographic orientation-dependent grain boundary properties, and the microstructure-dependent local elastic strain/strain energy density.

These effects have been identified through our recent model of whisker growth by grain boundary sliding limited creep (with and without coupling between grain boundary migration and shear stress), finite element analysis of a wide range of textured microstructures in Sn films, and recent experimental studies of microstructural evolution during intermetallic compound (IMC)-induced and thermal-cycling-induced stresses [11].

In addition, we have identified specific film microstructures and textures that are predicted to produce whisker-prone environments under specific stress conditions. By simulating the relationship between film texture (axis and strength) and the localization of the strain energy in the microstructure, a necessary condition for whisker formation is proposed: grains with locally high strain energy will become whiskers. Thus, measuring global film properties (crystallographic texture and strain energy density) may be a useful first step in identifying whisker-prone films, but is not sufficient in identifying which grain will form a surface defect.

Whisker growth will depend not only on local elastic strain energy density (ESED) but also grain geometry and the response of the grain boundaries surrounding the whisker grain to accretion-induced shear stresses.

1.2 CHARACTERISTICS OF WHISKER AND HILLOCK GROWTH FROM SURFACE GRAINS

Whiskers and hillocks typically grow from surface grains (Fig. 1.1) in fine-grained Sn and Sn-alloyed films in both isothermal (at room temperature or high temperature) and thermal cycling conditions. The origin of surface grains as nuclei for whiskers and hillocks have been much debated – whether surface grains were pre-existing

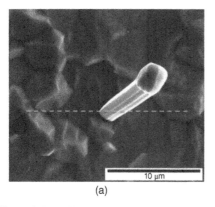 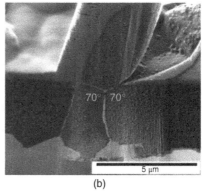

(a)   (b)

Figure 1.1 Whisker growing from surface grain. Image in (a) from secondary electron microscopy (SEM) shows a short whisker growing at an angle to the film top surface, while image in (b) from focused ion beam (FIB) cross section shows shallow grain boundaries (70° with respect to the film normal) at the whisker root. Source: Reprinted with permission from Elsevier © 2013 [12].

grains from the film deposition process [7, 13, 14], new grains that formed via surface recrystallization after the deposition process [7, 15, 16], or pre-existing grains whose topology changes as a result of grain growth in the film.

There is now direct evidence that whiskers and hillocks may form from either pre-existing or newly nucleated grains. Pei et al. [14] and Wang et al. [17] demonstrated that pre-existing grains could serve as nuclei for whiskers and hillocks for isothermal annealing (room temperature) and thermal cycling conditions, respectively.

Cross sections of whiskers formed under these conditions showed that the grain boundaries at the base of the whisker grain, that is, the "whisker root," are generally angled with respect to the film normal, creating shallow grains relative to the film thickness ("surface" grains) and that whiskers grew from grains with in-plane grain diameters typical of nonwhiskering grains in the film. In addition, surface grains that did not become whiskers or hillocks were also observed [12].

Shallow surface grains were observed by Sarobol et al. to form by recrystallization and grow into micron-sized whiskers and hillocks during thermal cycling of large-grained Sn-alloy films on Cu substrates, as seen in Figure 1.2 [18, 19]. New strain-free grains (low dislocation density) nucleated along pre-existing film grain boundaries and grew at the expense of highly deformed (high dislocation density) parent grains.

Hillocks are formed when growth out of the plane of the film is accompanied by grain boundary migration into the film, as seen in Figure 1.2c. Asymmetrical growth at the whisker root can lead to curved whiskers and hillocks, as shown in Figure 1.3 with a change in orientation of the whisker as it grows [8]. It is well known that the growth morphology for a given whisker and hillock may change over time, and since surface diffusion is suppressed by the presence of the surface oxide, the morphological changes over time are preserved in its surface shape [20].

1.2.1 Whisker Growth Model Based on Grain Boundary Sliding - Limited Creep

A growth model has been developed based on grain boundary faceting and grain boundary sliding limited Coble creep that explains the observed surface morphologies, the changing growth rate over time, and the dependence on grain geometry. In this model of whisker formation, two mechanisms are important: accretion of atoms by Coble creep on grain boundary planes normal to the growth direction (A_{accum}) inducing a grain boundary shear and grain boundary sliding on planes parallel to the direction of whisker growth (A_{slide}) with the grain boundary sliding coefficient determining whether accretion can lead to whisker formation.

The model accurately captures the importance of the geometry of surface grains – shallow grains on film surfaces whose depths are significantly less than their in-plane grain sizes.

A steady-state growth rate ($\Delta h/\Delta t$) of a whisker with constant radius can be calculated by

$$\frac{\Delta h}{\Delta t} = \frac{\Omega J A_{\text{accum}}}{\pi a^2} \qquad (1.1)$$

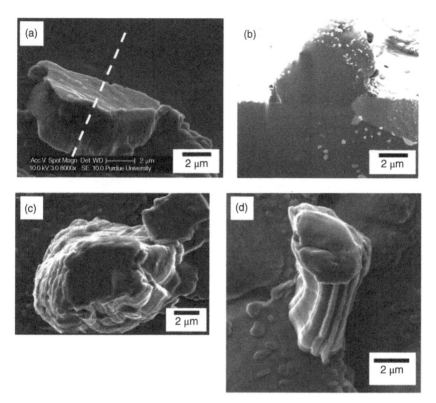

Figure 1.2   (a) Surface defect formed by recrystallization at a grain boundary, (b) with FIB cross section along the white dashed line in (a) showing shallow single-crystal grain with oblique grain boundaries, (c) representative hillock showing both surface uplift and grain boundary migration, and (d) representative whisker in thermally cycled solder film. Source: Reprinted with permission from Elsevier © 2013 [18].

where J is flux, A_{accum} is the area through which flux enters the surface grain and a is the surface grain radius. The flux, J, is estimated using a two-dimensional continuity equation, with $\nabla\sigma$ as the stress gradient that drives atomic flux toward the whisker base (σ_m at distance b away from the whisker center, which decreases toward the whisker edge). J and $\nabla\sigma$ are given by

$$J = \frac{D_{eff}\nabla\sigma}{RT} \tag{1.2}$$

$$\nabla^2\sigma = \frac{\partial^2\sigma}{\partial r^2} + \frac{1}{r}\frac{\partial\sigma}{\partial r} = 0 \tag{1.3}$$

$$\nabla\sigma|_{r=a} = \left.\frac{\partial\sigma}{\partial r}\right|_{r=a} = \frac{\Delta\sigma}{a\ln\left(\frac{b}{a}\right)} \tag{1.4}$$

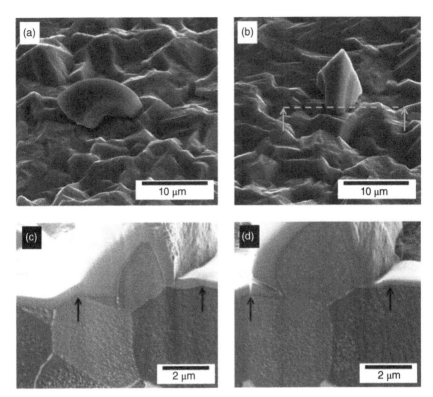

Figure 1.3 SEM images (a) and (b) from two different orientations of a curved whisker, which is decreasing in cross-sectional area as it grows. FIB images (c) and (d) show two successive cross sections of the whisker as oriented in (b), with the initial positions of the grain boundaries forming the whisker root indicated (black arrows). Source: Pei et al. [8].

The surface grain geometry is modeled as a stepped-cone shape with constant radius a and a cone angle of θ with effective accumulation area, $A_{accum} = \pi a^2$ and sliding area on the side, $A_{slide} = \frac{\pi a^2}{\tan \theta}$, as shown in Figure 1.4. A stress gradient creates a flux of atoms to the base of the surface grain. The accretion-induced shear stress has to overcome the grain boundary sliding friction, $\sigma_{friction}$. In the presence of an oxide layer, the accretion-induced shear stress must also crack the surface oxide (thickness, t, and strength τ), with $\sigma_{oxide} = \frac{2t}{a}\tau_{oxide}^{fracture}$, in order for grain boundary sliding to occur. The term $\Delta\sigma$ in Equation 1.4 is given by

$$\Delta\sigma = \sigma_m - \sigma_{friction}\frac{A_{slide}}{A_{accum}} - \sigma_{oxide} \qquad (1.5)$$

Thus, a back stress in the whisker grain can be created, stopping the growth if the accretion-induced shear is not high enough for sliding. In summary, the critical factors in the growth rate analysis are $\sigma_{friction}$, the in-plane film average compressive stress

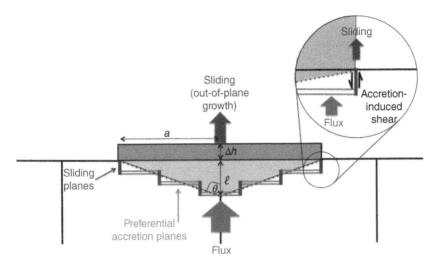

Figure 1.4 Schematics of a simple whisker growth model, delineating the roles of atomic accretion, faceting of the grain boundary (facets shown by vertical lines and horizontal double lines), and grain boundary sliding (along the sliding planes shown by vertical lines) on surface defect formation from a surface grain with ($0° < \theta < 90°$). Source: Reprinted with permission from Elsevier © 2013 [12].

σ_m, the stress required to produce a circumferential crack in the oxide of thickness t, and the surface grain geometry, a and θ. Thus, the steady-state growth rate of a surface grain as a whisker, with constant radius, derived from the two-dimensional continuity equation is

$$\frac{\Delta h}{\Delta t} = \left(\frac{\Omega D_{\text{eff}}}{RT \ln \left(\frac{b}{a} \right)} \right) \left(\frac{1}{a} \right) \left[\sigma_m - \left(\frac{1}{\tan \theta} \right) \sigma_{\text{friction}} - 2 \left(\frac{t}{a} \right) \tau_{\text{oxide}}^{\text{fracture}} \right] \qquad (1.6)$$

The implication of this steady-state growth model is that there exists a critical surface grain geometry for a given grain boundary sliding coefficient and oxide layer thickness, which determines if such a surface grain will grow out of plane via grain boundary sliding and Coble creep and, therefore, will become a whisker. Assuming that the effect of surface oxide thickness is relatively small, a normalized growth rate can be calculated as a function of surface grain depth and the $\sigma_{\text{friction}}/\sigma_m$ as shown in Figure 1.5.

The "No Growth" region in Figure 1.5 corresponds to the situation where the surface grain is deep (low θ) and the ratio between grain boundary sliding friction and the film stress ($\sigma_{\text{friction}}/\sigma_m$) is high. The "Growth" region in Figure 1.5 corresponds to the situation where the surface grain is shallow (θ approaches 90°) and the ratio between grain boundary sliding friction and the film stress ($\sigma_{\text{friction}}/\sigma_m$) is low. Cross sections of whisker grains consistently showed that the surface grains were shallow

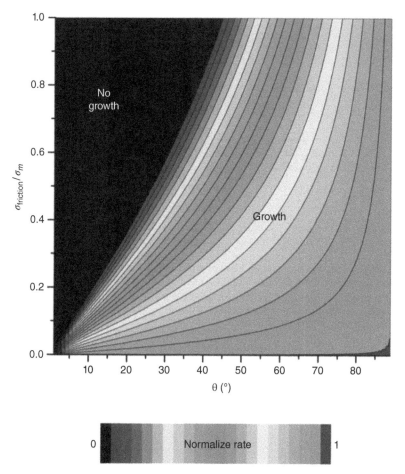

Figure 1.5   Effects of surface grain geometry and grain boundary sliding on whisker growth rate. The growth rate is normalized by the maximum growth rate and projected as a contour on the plot as a function of θ and $(\sigma_{friction}/\sigma_m)$. The critical θ_c for zero growth rate increases as the $(\sigma_{friction}/\sigma_m)$ increases, approaching $45°$ when $(\sigma_{friction}/\sigma_m) = 1$. The region of interest is $60° < \theta < 90°$, corresponding to those observed from cross sections of whiskers.   Source: Reprinted with permission from Elsevier © 2013 [12]. Please visit www.wiley.com/go/Kato/TinWhiskerRisks to access the color version of this figure.

with $\theta > 60°$. Based on this model, long whiskers grow from shallow surface grains with easy grain boundary sliding in the direction of growth.

During growth by sliding limited creep without coupling between the shear stress and grain boundary migration, the grain boundaries forming the whisker root will remain stationary, resulting in a whisker with a constant diameter. In the case where growth is coupled with shear-induced grain boundary migration, the grain boundaries around a particular whisker may either move toward the center of the whisker, resulting in a whisker with a decreasing diameter and changing grain boundary curvatures to maintain the force balances at the interface junctions

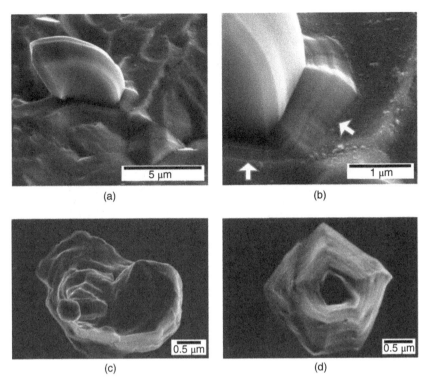

Figure 1.6   SEM images of (a) a whisker with a decreasing diameter as it grows, (b) higher magnification view of the base of the whisker shows the initial position of the surface grain (indicated by arrows) and surface striations corresponding to the positions of the grain boundary groove between the whisker and neighboring grain as it shrank, (c) whisker-to-hillock transition, and (d) hillock formation.

(Figs. 1.3 and 1.6a, b), or move away from the center of the whisker, resulting in a hillock (Fig. 1.6d) or whisker-to-hillock transition (Fig. 1.6c).

The occurrence of grain boundary migration and its directionality during whisker and hillock formation may be an indication of coupling between accretion-induced shear stress and grain boundary migration. The conditions and mechanisms for coupling migration and applied shear stresses are being elucidated through molecular dynamics (MD) and phase-field crystal simulations (PFC) [21, 22].

Cahn et al. [21] identified a wide range of processes that may result from coupling, including a transition between coupling modes that may lead to a stationary grain boundary and a "process that will look like grain boundary sliding." They proposed that many phenomena, including diffusion-induced grain boundary migration (DIGM), may be caused by such coupling. Whisker morphologies such as the one shown in Figure 1.7 suggest that the lack of coupling between the accretion-induced shear stress and grain boundary migration may lead to the formation of long, straight whiskers, the ones more likely to cause short circuits in electronics.

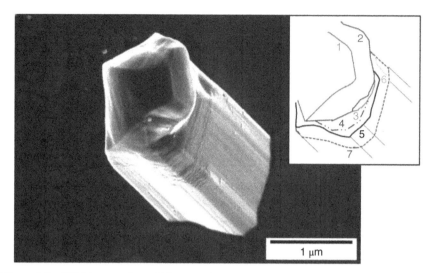

Figure 1.7   SEM image of a whisker growing on an electroplated Sn film, with the grain boundary contours (inset) of the whisker showing the changes in diameter and growth symmetry over time. Formation of a straight whisker seems to be associated with the cessation of grain boundary migration.

In Figure 1.7, the progression of grain boundary positions as the whisker grows can be seen starting from the tip of the whisker, where contour 1 corresponds to the original surface of the whisker. Contours 2 through 7 mark the positions of the grain boundaries between the whisker and its neighboring grains over time, with the whisker growing asymmetrically as it first increases, then decreases, and then increases in diameter.

Starting at approximately contour 6, the whisker grows with a close to constant profile, that is, the grain boundaries surrounding the whisker are no longer migrating as the whisker grows out of the film plane. Extensive, systematic MD and PFC studies of surface grains with different grain boundary orientations are needed to help us better understand the role of coupling and grain boundary sliding in whisker and hillock formation.

1.2.2   Whisker Formation and Fatigue Damage Accumulation during Thermal Cycling

The grain boundary sliding limited model for whisker growth explains whisker growth behavior not only during intermetallic growth at room temperature but also under thermal cycling conditions, where we have observed that the whisker radius changes with the number of thermal cycles [17].

Images of individual whiskers as a function of thermal cycling (−40 to 85°C) revealed that whiskers grew and their radii decreased, with a deep groove at the whisker roots, and eventually "pinched off." This whisker "pinch-off" process led to a decreasing whisker density with increasing number of thermal cycles.

Figure 1.8 SEM images of a whisker at 20 and 50 cycles. The whisker on the left pinched off and disappeared between 20 and 50 cycles, leaving behind a remnant hole.

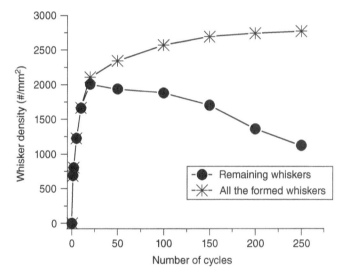

Figure 1.9 Whisker density as a function of number of thermal cycles: the circle symbol corresponds to the density of whiskers that remained on the sample surface; the star symbol corresponds to the density of all the formed whiskers, including those that disappeared after 20 cycles.

An example of this whisker "pinch-off" phenomenon is shown in Figure 1.8, where the whisker (with decreasing radius) disappeared between 20 and 50 thermal cycles and a new whisker started to grow. While the total number of whiskers (counting both the intact whiskers and the remaining holes of the whiskers that were "pinched off") increased with increasing thermal cycles, the rate of whisker "pinch-off" was much more rapid than the rate of whisker nucleation (Fig. 1.9).

The mechanistic model Wang et al. [17] proposed (as an extension to the model described earlier) for the pinch-off process is summarized in Figure 1.10.

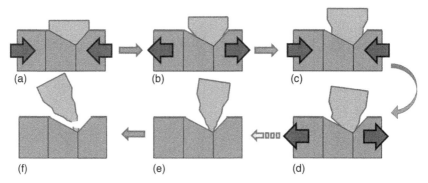

Figure 1.10   Schematic of the whisker pinch-off phenomenon. (a) Whisker grows due to a compressive stress upon heating; (b) during cooling, a crack forms at the whisker root due to the tensile stress; (c) subsequent heating leads to whisker growth with a smaller diameter; (d) crack propagates along whisker grain boundaries due to a tensile stress upon cooling; (e) whisker continues to grow with a further decreased diameter; and (f) when the whisker diameter is small enough, it becomes fragile enough to break off during handling. Source: Wang et al. [17].

Due to the coefficient of thermal expansion (CTE) mismatch between the Sn film and the Si substrate, the Sn film is under compression upon heating and under tension upon cooling. During the heating cycle, the whisker grows out of the film plane via grain boundary sliding to relieve the compressive stress. During the cooling cycle, tensile stress induces a crack at the whisker root. With the newly exposed Sn at the crack being oxidized, the crack cannot close when the tensile stress is removed and the growth during the next heating cycle leads to horizontal striation rings on the whisker surface. One horizontal striation ring corresponds to whisker growth during one thermal cycle [23].

With each subsequent heating and cooling cycle, the crack propagates down the oblique whisker grain boundaries, creating a deep groove at the whisker root, as seen in Figure 1.8. This groove causes a decrease in whisker radius with each thermal cycle, until the ever more fragile whiskers break off during routine handling of the films, that is, "pinch-off." Striations are seen along both the whisker and the groove surfaces in the film that correspond to individual thermal cycles.

The implication of the "pinch-off" on the whisker growth rate is significant. As the whisker radius, a, decreases in the initial stage of the pinch-off process, the flux, J, to the base of the whisker increases, according to Equations 1.2 and 1.4. However, as the whisker diameter decreases, the growth of the whisker will, at some diameter, become kinetic-limited, either by diffusion along the whisker root or interface attachment.

As a decreases, grain boundary cracking and groove formation are often observed to be asymmetrical, with the whisker becoming completely detached from some neighboring grains. The resulting decrease in grain boundary area leads to an overall decrease in J. At this point, the whisker growth rate is better described by a one-dimensional continuity equation, where the stress gradient (i.e., the driving force for growth) and thus J decrease with decreasing a. The linearized stress gradient is given by

$$\nabla \sigma|_{x=a} = \frac{\partial \sigma}{\partial x}\bigg|_{x=a} = \frac{\Delta \sigma}{3a_0 - a} \tag{1.7}$$

where a_0 is the original radius of the surface grain before the "pinch-off" process occurred.

In the case of a continuously reforming surface oxide, the accretion-induced shear stress may no longer be sufficient to break the oxide, which increases as a decreases, that is, $\sigma_{oxide} = 2(t/a)^* \tau_{oxide}$, such that

$$\sigma_m < \sigma_{friction}\frac{A_{slide}}{A_{accum}} + \sigma_{oxide} \tag{1.8}$$

Therefore, with increasing thermal cycles, the whisker growth rate initially increases and later decreases either until whisker "pinch-off" occurs or until growth stops when the accumulation-induced shear force is insufficient to overcome the grain boundary sliding friction or break the oxide.

Although the grain boundary sliding limited creep model provides some insights into the roles of grain geometry and grain boundary sliding in whisker growth, the microstructural selection process that predisposes some shallow surface grains to become whiskers and hillocks must be refined further. In the next section, the effects of grain boundary sliding friction and localized stress, which are highly dependent on grain boundary structure, grain orientations, and misorientations, are discussed.

1.2.3 Preferred Site Model Based on Crystallographic Misorientations as well as Elastic and Thermoelastic Anisotropic Properties

Besides the surface grain geometry and grain boundary sliding friction, there are several additional factors that predispose a particular surface grain to becoming a whisker. These factors include localization of high strain energy (the driving force for whisker formation) due to variation in local crystallographic orientations. Object-oriented finite element analysis (OOF) [24, 25] was utilized to examine the generation of local high ESEDs for the Sn grains with specific misorientations from the neighboring grains [26].

Grain orientations generated using the March–Dollase orientation distribution function [27, 28] for (001), (100), (110), and (111) fiber textures were randomly assigned to the grains in a two-dimensional, polycrystalline microstructure [where (hkl) planes of the grains are preferentially parallel to the film plane]. The plane stress condition ($\sigma_{zz} = \sigma_{xz} = \sigma_{yz} = 0$) was imposed to simplify the calculation for thin films. Elastic stresses induced at room temperature due to Cu_6Sn_5 IMC formation were generated assuming an in-plane biaxial compressive strain ($\varepsilon_1 = \varepsilon_2 = -0.01\%$) imposed on the films. For thermoelastic stresses during thermal cycling, the CTE mismatch between the Sn film and the Cu substrate was simulated by constraining the in-plane film strain to match the thermal expansion of the substrate for a

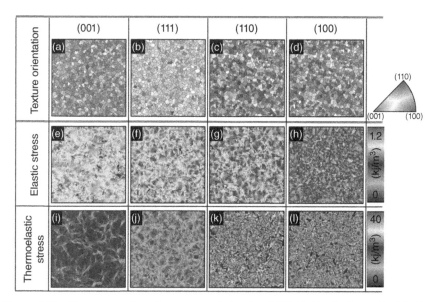

Figure 1.11 The grain orientation maps and simulated ESED distributions. (a)–(d) Normal direction grain orientation maps in inverse pole figure (IPF) colors for (001), (111), (110), and (100) textured films, respectively. The corresponding ESED distributions (e)–(h) for elastic stresses induced by room temperature aging due to IMC formation, and (i)–(l) for thermoelastic stresses induced during thermal cycling. ESED distributions for the same stress-generating mechanisms are scaled in the same manner. Please visit www.wiley.com/go/Kato/TinWhiskerRisks to access the color version of this figure.

temperature change from 25 to 85°C. The global and local responses to these stress conditions were analyzed in terms of ESEDs.

The global response varies significantly with the film texture, as shown in Figure 1.11. The textures for films with the lowest overall ESED are (100) textures for room temperature aging and (001) textures for thermal cycling conditions. In addition to the overall ESED for the entire film, locally high ESEDs associated with specific grain misorientations were observed and are shown in Figure 1.12.

These high ESED grains are identified as potential whiskers. The orientations of all the grains and those with high ESEDs (top 1%) are plotted using normal direction inverse pole figures (ND IPFs) in Figure 1.12.

For room temperature aging, (001) and (111) textured films have high ESED grains mostly orientated toward the (100). The orientations of high ESED grains evidently differ significantly from the primary texture. On the other hand, (110) textured films have most high ESED grains oriented toward the (110) and some oriented toward the (100), whereas (100) textured films have high ESED primarily grains oriented toward the (100), identical to the primary texture.

For thermal cycling condition, (001) and (111) textured films have high ESED grains oriented between (110) and (100), far away from the texture, but the (110) and (100) textured films have high ESED grains with orientations similar to the textures.

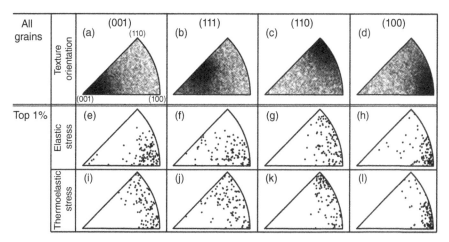

Figure 1.12   ND IPFs showing grain orientation distributions (a)–(d) for (001), (111), (110), and (100) textured films, respectively. Orientations of grains with high ESEDs (top 1%) for each texture are shown in ND IPFs (e)–(h) for elastic stresses and (i)–(l) for thermoelastic stresses. Each dot stands for one grain orientation.

In summary, the anisotropic elasticity (effective in-plane biaxial moduli) for room temperature aging and thermoelasticity (CTE mismatch or thermal cycling) caused nonuniform, highly localized ESEDs. These high ESED grains commonly have their c-axes aligned with the film plane but also depend on the textures and stress conditions. Thus, further analysis must take into account all the directions in orientation space (ND and in-plane directions), as well as the anisotropic elastic and thermoelastic properties of β-Sn.

For the case of a film with a (111) texture, the local elastic strains have been measured and compared to the simulation results for the room temperature aging condition. Local grain orientations, deviatoric elastic strains, and ESEDs have been measured using synchrotron microdiffraction in regions containing whiskers or hillocks for films, whose primary texture is near (111) [11].

An example of the comparison is shown in Figure 1.13: a secondary electron microscopy (SEM) image of a scanned area around a whisker (Fig. 1.13a), the orientations of the whisker grain (811) and the other grains in the film projected in normal direction (ND) inverse pole figure (IPF) (Fig. 1.13b), the corresponding microdiffraction (Fig. 1.13c), and simulated (Fig. 1.13d) grain orientation maps, measured out-of-plane elastic strain (Fig. 1.13e) and ESED distribution maps (Fig. 1.13g), as well as simulated out-of-plane elastic strain (Fig. 1.13f) and ESED distribution (Fig. 1.13h) maps. Similarly for two other whiskers and six hillocks in the room temperature aging condition, the whisker/hillock grain orientations differ significantly from the film textures.

The three whisker and six hillock grains were observed to have higher crystallographic misorientations with neighboring grains (average peak misorientation angle of 65° for whiskers and 80° for hillocks) than generally observed in the

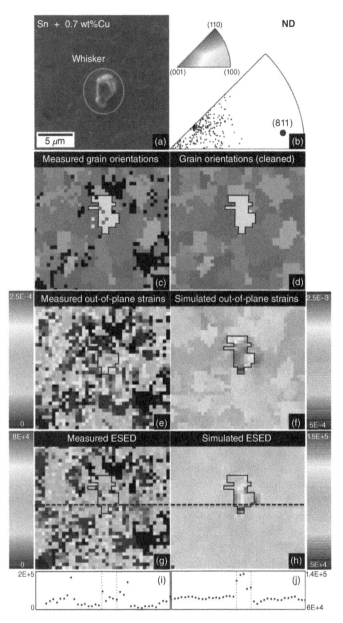

Figure 1.13   (a) SEM image of a whisker on Sn + 0.7 wt%Cu film. (b) Orientation of the whisker grain (pink), (811), and other grains in the film (black) displayed in ND equal-area IPF. (c) As-collected grain map in normal direction inverse pole figure colors. (d) Cleaned grain map in ND IPF colors. (e) Measured out-of-plane strain distribution from microdiffraction. (f) OOF-simulated out-of-plane strain distribution. (g) Measured ESED distribution. (h) OOF-simulated ESED distribution. (i) and (j) are line plots of measured and simulated ESED as a function of distance across the whisker grain corresponding to the position of the dashed line in (g) and (h).   Source: Reprinted with permission from Cambridge University Press © 2013 [11]. Please visit www.wiley.com/go/Kato/TinWhiskerRisks to access the color version of this figure.

microstructure (average peak misorientation angle of 50° for randomly selected grain pairs). While elastic simulations predicted higher local out-of-plane elastic strains and ESEDs for whisker and hillock grains, synchrotron measurements of out-of-plane strains of whisker and hillock grains after growth showed relaxation, with correspondingly low ESEDs calculated from measured strains.

This comparison suggests that, before whisker or hillock formation, highly misoriented grains with high out-of-plane elastic strains and ESEDs relative to their neighbors determined, at least in part, which grains became whiskers or hillocks. Due to the highly anisotropic properties of β-Sn, the local grain orientations as well as misorientations determine grains with high ESEDs, in response to external strains (due to IMC formation or temperature changes).

The combined results from simulation and experiment verify that grains with high ESEDs are potential whisker formation sites. Extending this to the overall microstructure and to the films, we have identified whisker-resistant textures (low ESEDs) under different stress-generating mechanisms: a strong (100) fiber texture film for a compressive stress induced only by IMC formation and a strong (001) fiber texture film for a compressive stress induced during thermal cycling.

1.3 SUMMARY AND RECOMMENDATIONS

In this chapter, we examined the characteristic properties of the microstructure around whisker and hillock grains with the aim to further identify why these particular shallow surface grains and locations became predisposed to forming whiskers and hillocks.

On the local level, three major factors play a role in determining which surface grains will become a whisker/hillock, including the surface grain geometry, crystallographic orientation-dependent surface grain boundary structure, and the localization of elastic strain/strain energy density distribution due to variation in crystallographic orientations and β-Sn anisotropy.

A growth model was developed, based on grain boundary faceting, localized Coble creep, and grain boundary sliding for whiskers. In this model of whisker formation, two mechanisms are important: accretion of atoms by Coble creep on grain boundary planes normal to the growth direction inducing a grain boundary shear and grain boundary sliding in the direction of whisker growth.

The model accurately captures the importance of the geometry of surface grains – shallow grains on film surfaces whose depths are significantly less than their in-plane grain sizes. On the basis of this model, long whiskers grow from shallow surface grains with easy grain boundary sliding in the direction of growth.

For the thermal cycling condition, a whisker "pinch-off" phenomenon was observed, where the whisker radius decreases as growth occurs. The decrease in whisker radius resulted in the initial increase in the flux (and thus the growth rate), followed by the decrease in stress gradient and increase in resistance to oxide breaking, resulting in reduced growth rate.

Local grain orientations and strains measured by synchrotron microdiffraction in regions containing whiskers or hillocks were compared with elastic finite element analysis simulations. Whisker and hillock grains were observed to have higher crystallographic misorientations with neighboring grains than generally observed in the microstructure.

While elastic simulations predicted higher local out-of-plane elastic strains and ESEDs for whisker and hillock grains, synchrotron measurements of out-of-plane strains of whisker and hillock grains after growth showed relaxation, with correspondingly low ESEDs calculated from measured strains. This suggests that, before whisker or hillock formation, highly misoriented grains with high out-of-plane elastic strains and ESEDs relative to their neighbors determined, at least in part, which grains became whiskers or hillocks.

These two-dimensional simulations of film response to strains imposed by two different aging conditions – room temperature and thermal cycling – may therefore be a useful method for evaluating and predicting the propensity of a film with a particular texture to form whiskers. Hence, it may be possible to engineer a whisker-resistant texture for the appropriate aging condition. Furthermore, while complete elimination of surface grains may be impossible, the combination of a whisker-resistant texture and a uniformly high grain boundary sliding coefficient may serve to significantly reduce the risk of whisker formation.

ACKNOWLEDGMENTS

The authors gratefully acknowledge support from NSF Graduate Research Fellowship Program, Cisco Systems, Inc., Foresite Inc., Army Research Laboratories, and Naval Surface Warfare Center (Crane Division). We would like to thank Dr Peng Su, Maureen Williams, Dr Anthony Rollett, Dr Martin Kunz, and Dr Nobumichi Tamura for assistance with data extraction and valuable discussions.

We acknowledge support from DOE-BES (DE-FG02-05ER15637) and NSF (EAR-0337006) and access to ALS beamline 12.3.2. ALS is supported by the Director, Office of Science, Office of Basic Energy Sciences, Materials Science Division, of the US Department of Energy under Contract no. DE-AC02-05CH11231. The microdiffraction program at the ALS beamline 12.3.2 was made possible by NSF grant # 0416243.

Sandia National Laboratories is a multiprogram laboratory managed and operated by Sandia Corporation, a wholly owned subsidiary of Lockheed Martin Corporation, for the US Department of Energy's National Nuclear Security Administration under contract DE-AC04-94AL85000.

This chapter is reproduced from Springer Journal of Metals, A Predictive Model for Whisker Formation Based on Local Microstructure and Grain Boundary Properties, 65 [10] (2013) 1350–1361, P. Sarobol, Y. Wang, W. H. Chen, A.E. Pedigo, J.P. Koppes, J.E. Blendell, and C.A. Handwerker, published online August 24, 2013, with kind permission of Springer Science+Business Media.

REFERENCES

1. Sobiech M, Welzel U, Mittemeijer EJ, Hügel W, Seekamp A. Driving force for Sn whisker growth in the system Cu–Sn. Appl Phys Lett 2008;93:011906.

2. Sobiech M, Wohlschlögel M, Welzel U, Mittemeijer EJ, Hügel W, Seekamp A, Liu W, Ice GE. Local, submicron, strain gradients as the cause of Sn whisker growth. Appl Phys Lett 2009;94:221901.

3. Tu KN. Irreversible processes of spontaneous whisker growth in bimetallic Cu–Sn thin-film reactions. Phys Rev B 1994;49:2030.

4. Lee BZ, Lee DN. Spontaneous growth mechanism of tin whiskers. Acta Mater 1998;46:3701.

5. Chason E, Jadhav N, Chan WL, Reinbold L, Kumar KS. Whisker formation in Sn and Pb–Sn costings: role of intermetallic growth, stress evolution and plastic deformation process. Appl Phys Lett 2008;92:171901.

6. Kumar KS, Reinbold L, Bower A, Chason E. Plastic deformation processes in Cu/Sn bimetallic films. J Mater Res 2008;23:2916.

7. Boettinger WJ, Johnson CE, Bendersky LA, Moon K-W, Williams ME, Stafford GR. Whisker and Hillock formation on Sn, Sn–Cu and Sn–Pb electrodeposits. Acta Mater 2005;53:5033.

8. Pei F, Jadhav N, Chason E. Correlation between surface morphology evolution and grain structure: Whisker/hillock formation in Sn–Cu. JOM 2012;64:1176.

9. Moon KW, Johnson CE, Williams ME, Kongstein O, Stafford GR, Handwerker CA, Boettinger WJ. Observed correlation of Sn oxide film to Sn whisker growth in Sn–Cu electrodeposit for Pb-free solders. J Electron Mater 2005;34:L31.

10. Jadhav N, Buchovecky E, Chason E, Bower A. Real-time SEM/FIB studies of whisker growth and surface modification. JOM 2010;62:30.

11. Sarobol P, Chen WH, Pedigo AE, Su P, Blendell JE, Handwerker CA. Effects of local grain misorientation and beta-Sn elastic anisotropy on whisker and hillock formation. J Mater Res 2013;28:747.

12. Sarobol P, Blendell JE, Handwerker CA. Whisker and hillock growth via coupled localized Coble creep, grain boundary sliding, and shear induced grain boundary migration. Acta Mater 2013;61:1991.

13. Frolov T, Boettinger WJ, Mishin Y. Atomistic simulation of hillock growth. Acta Mater 2010;58:5471.

14. Pei F, Jadhav N, Chason E. Correlating whisker growth and grain structure on Sn–Cu samples by real-time scanning electron microscopy and backscattering diffraction characterization. Appl Phys Lett 2012;100:221902.

15. Galyon GT, Palmer L. An integrated theory of whisker formation: the physical metallurgy of whisker formation and the role of internal stresses. IEEE Trans Electron Packag Manuf 2005;28:17.

16. Vianco PT, Rejent JA. Dynamic recrystallization (DRX) as the mechanism for Sn whisker development. Part I: A model. J Electron Mater 2009;38:1815–1825.

17. Wang Y, Blendell JE, Handwerker CA. Evolution of tin whiskers and subsiding grains in thermal cycling. J Mater Sci 2014;49:1099–1113.

18. Sarobol P, Koppes JP, Chen WH, Su P, Blendell JE, Handwerker CA. Recrystallization as a nucleation mechanism for whiskers and hillocks on thermally cycled Sn-alloy solder films. Mater Lett 2013;99:76.

19. Koppes JP, Ph.D. Thesis, Purdue University: West Lafayette, IN; 2012.

20. Pedigo AE, Handwerker CA, Blendell JE, Proceedings of the Electronic Components and Technology Conference; Lake Buena Vista, Florida, USA: IEEE; 2008. p 1498.

21. Cahn JW, Mishin Y, Suzuki A. Coupling grain boundary motion to shear deformation. Acta Mater 2006;54:4953.

22. Trautt ZT, Adland A, Karma A, Mishin Y. Coupled motion of asymmetrical tilt grain boundaries: molecular dynamics and phase field crystal simulations. Acta Mater 2012;60:6528.

23. Suganuma K, Baated A, Kim K-S, Hamasaki K, Nemoto N, Nakagawa T, Yamada T. Sn whisker growth during thermal cycling. Acta Mater 2011;59:7255.

24. Langer SA, Fuller E, Carter WC. OOF: an image-based finite-element analysis of material microstructures. Comput Sci Eng 2001;3:15.

25. Reid ACE, Lua RC, García RE, Coffman VR, Langer SA. Modelling microstructures with OOF2. Int J Mater Prod Technol 2009;35:361.

26. Chen WH, Sarobol P, Holaday J, Handwerker CA, Blendell JE. Effect of crystallographic texture, anisotropic elasticity and thermal expansion on whisker formation in β-Sn thin films. J Mater Res 2014;29:197–206.

27. Blendell JE, Vaudin MD, Fuller ER. Determination of texture from individual grain orientation measurements. J Am Ceram Soc 1999;82:3217.

28. Dollase WA. Correction of intensities for preferred orientation in powder diffractometry: application of the March model. J Appl Crystallogr 1986;19:267.

2

MAJOR DRIVING FORCES AND GROWTH MECHANISMS FOR TIN WHISKERS

ERIC CHASON
Divison of Engineering, Brown University, Providence, Rhode Island, USA

NITIN JADHAV
IBM, Hopewell Junction, New York, USA

2.1 INTRODUCTION

The focus of this chapter is on describing the primary driving forces and mechanisms causing the growth of whiskers and hillocks. The impact of Sn whiskers on reliability and the importance of suppressing their formation are discussed in detail in other sections of this book, so we do not review them here. In particular, we describe the results of experiments that have been designed to distinguish among different potential mechanisms and identify the most important ones so as to develop a fuller understanding of the materials science behind whisker growth. This understanding is needed for the development of mitigation strategies that can get to the root causes of whisker formation. Moreover, knowing the mechanisms of growth enables us to develop models that can predict whisker formation and interpret results of accelerated aging studies.

It has long been suspected that the primary driving force for whisker formation is stress. As early as 1954, it was shown that the application of mechanical stress to

Mitigating Tin Whisker Risks: Theory and Practice, First Edition.
Edited by Takahiko Kato, Carol A. Handwerker, and Jasbir Bath.
© 2016 John Wiley & Sons, Inc. Published 2016 by John Wiley & Sons, Inc.
Companion website: www.wiley.com/go/Kato/TinWhiskerRisks

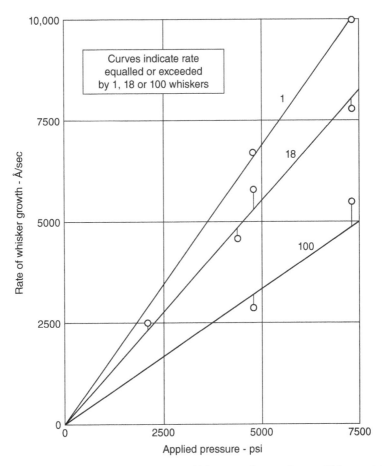

Figure 2.1   Effect of applied pressure on whisker growth rate. Source: Fisher et al. [1].

Sn layers resulted in the growth of whiskers from the surface [1, 2]. A plot from this work (Fig. 2.1) of growth rate versus applied pressure shows that the whisker growth rate is accelerated by the application of mechanical stress. Other studies [3–6] have repeatedly confirmed that mechanical stress can enhance whisker growth. Oxidation has also been proposed as an alternative driving force, but we believe it can be discounted as the primary cause by experiments that show that whiskers still form under high vacuum conditions where oxidation will not occur [7].

Of course, if applied mechanical forces were the only reason of whiskering, then they could easily be suppressed by eliminating the origin of the stress (clamping, bending, etc.). Other sources of stress that have been identified are not as easily removed; these include residual stress [6] due to the initial plating [8], thermal-expansion mismatch during thermal cycling [9], electromigration and intermetallic compound (IMC) formation between the Sn and Cu layers [6, 10–12].

It has also been suggested that whisker enhancement during corrosion is due to stress induced by the formation of corrosion products [13]. The stresses that are a one-time occurrence (e.g., mechanical or residual stress) are eventually removed by stress relaxation processes in the layer. For instance, it was shown by Pitt and Henning [14] that the whisker growth slowed down as the applied pressure was relieved by deformation of the Sn. The more problematic stress sources are the ones that are applied repeatedly (thermal cycling, electromigration) or get replenished over time (IMC formation) since they are not eliminated by stress relaxation processes.

In this work, we concentrate primarily on the effects of IMC formation because it is ubiquitous in Sn platings on Cu conductors. Numerous measurements [6, 10–12, 15–18] have shown that IMC formation leads to stress in the Sn layer. Lee and Lee [11] were the first to show this quantitatively by using wafer curvature to measure the buildup of compressive stress in the Sn layer. Tu [10] noted that Cu diffuses more rapidly into Sn than Sn does into Cu. This leads to the formation of IMC preferentially on the Sn side of the Sn–Cu interface. The increase in the volume of material on the Sn side corresponds to a volumetric strain that is the source of the stress in the Sn layer.

However, correlation of stress with whisker formation is not sufficient to determine the mechanisms that lead to the stress evolution and the whisker formation. We, therefore, describe the results of experimental studies that have been designed to elucidate some of the mechanisms behind the stress evolution and the resulting whisker and hillock formation. Although the experiments focus on IMC-induced stress, the mechanisms of stress relaxation and surface evolution are relevant to whisker formation due to other causes as well.

The first part of this chapter is devoted to describing how the growth of IMC leads to stress in the Sn layer. We discuss results from simultaneously measuring the growth of the intermetallic phase (IMC) between the Cu and Sn whiskers, the stress in the Sn layer, and the whisker density to understand how these different processes interact. By systematically modifying the Sn structure, we are able to identify which aspects of the layer are responsible for controlling the stress evolution and hence the whisker formation. We also describe experimental studies that are designed to identify the other processes that play a role in whiskers forming, for example, plasticity, diffusion-mediated creep, and prevention of surface diffusion by the Sn oxide. The mechanisms determined from these experiments are combined into a finite-element analysis (FEA) simulation to show how the stress induced by the formation of the IMC leads to stress throughout the layer.

In the second part, we describe results from real-time studies of whisker growth in the scanning electron microscope/focused ion beam milling (SEM/FIB) that show how whiskers and hillocks evolve in time. FIB modification of the surface is used to confirm that the underlying microstructure determines where whiskers will form, not just the weakness of the surface oxide, indicating that there are certain grains with a propensity to grow into whiskers and hillocks. Real-time videos show the actual process of whisker/hillock growth and indicate that there is a complex interplay between diffusion of material to their base, plastic deformation in the growing feature, and grain growth into the surrounding Sn matrix.

2.2   UNDERSTANDING THE MECHANISMS BEHIND IMC-INDUCED STRESS EVOLUTION AND WHISKER GROWTH

It is known that many features of the Sn layers (plating bath, microstructure, grain size, texture, etc.) play a strong role in determining whisker formation, so in order to determine underlying mechanisms, it is necessary to carefully control and characterize the films. As discussed next, for instance, the grain size can greatly influence the rate of IMC growth or stress relaxation. If the grain size is not measured or reported, then it is difficult to compare the results from films that are grown by different methods and with different structures. This can make comparison between studies reported in the literature problematic and their meaning hard to interpret. To circumvent this problem, in the studies described next, we have used a series of samples, which have been systematically modified to look at effects of factors such as grain size, thickness, composition, and surface oxide on the evolution of the IMC volume, stress, and whisker density. The samples were grown by electrodeposition using commercial plating solutions for pure Sn and Pb–Sn alloys. The Sn-based layers were grown over Cu layers that were vapor deposited via electron-beam deposition on oxidized Si (100) substrates. Further details of the deposition process can be found in [6, 18].

Before describing the experimental results, it is important to understand the multiple layers that make up the structure as illustrated by the cross section of a whisker growing out of an Sn surface shown in Figure 2.2. In this case, the Sn layer has a columnar microstructure with a grain size comparable to its thickness (which is not always the case for different growth conditions or electroplating solutions). The Cu layer has a much finer grain size (100 nm) in which the grains are relatively equiaxed with a smooth interfacial morphology before Sn deposition. The IMC is seen to grow primarily into the Sn–Sn grain boundaries at the Sn–Cu interface, consistent with observations by others [15–17]. Overlying the Sn layer is a tenacious native oxide with a thickness of approximately 5 nm. Although the image in Figure 2.2 shows a grain that developed into a whisker-like feature, such an occurrence is relatively rare (i.e., on this sample only roughly 1 out of 10^3–10^4 grains shows any deformation into whiskers or hillocks).

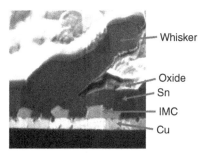

Figure 2.2   FIB cross section of the Sn layer plated over Cu. The arrows indicate the presence of whisker, oxide layer, Sn, IMC, and Cu, respectively. Source: Chason et al. [19].

2.2.1 Simultaneous Measurement of IMC Growth, Stress Evolution, and Whisker Density

In this section, we describe experiments to look at the simultaneous evolution of IMC volume, stress, and whisker density. As discussed next, the measured correlation between these quantities on a series of samples with different thickness, grain size, and so on enables us to identify specific mechanisms that are involved in controlling the stress evolution and whisker formation.

Measurements from a set of samples with different Sn layer thickness (1.45, 2.9, and 5.8 µm) are shown in Figure 2.3. A schematic of the film structure is shown at the top of the figure (Fig. 2.3a–c). The next row (Fig. 2.3d–f) shows an SEM image of the Cu–Sn interface after the Sn has been removed by selective etching (6 h after the film was deposited). Removing the Sn reveals the growth of the IMC into the Sn–Sn grain boundaries (bright spots in images) and leaves a marker for the Sn grain structure. The next panels show (g–i) the total volume of the IMC that has formed, (j–l) the corresponding average stress in the Sn layer, and (m–o) the whisker density.

The measurements of IMC volume were obtained by measuring the change in weight [18, 20, 21] when the sample is selectively etched to remove the unreacted Sn (without affecting the IMC or underlying Cu). Comparison with the weight of the Sn layer measured at the time of deposition corresponds to the amount that has been incorporated into the Cu_6Sn_5 IMC. The SEM measurements of the surface after the Sn is removed enable us to determine the Sn grain size (Fig. 2.3d–f) since the IMC grows preferentially along the Sn grain boundaries.

The stress in the Sn layer is measured using a multibeam optical stress sensor (MOSS) wafer curvature technique [22] on the same samples that are used to determine the IMC volume. For a multilayer structure, the curvature of the substrate induced by the film stress is related to the product of layer stress and thickness summed over all the layers in the structure [23]:

$$\frac{1}{R} = \frac{6}{M_s h_s^2}[\langle\sigma_{Cu}\rangle h_{Cu} + \langle\sigma_{IMC}\rangle h_{IMC} + \langle\sigma_{Sn}\rangle h_{Sn}] \tag{2.1}$$

$\langle\sigma_i\rangle$ is the average stress and h_i is the thickness of each layer of the film (Cu, IMC, and Sn) so that $\langle\sigma_i\rangle h_i$ is equal to the in-plane stress integrated over the thickness of each layer. M_s and h_s refer to the biaxial modulus and thickness of the substrate, respectively. Note that the curvature is due to multiple layers, so the stress of each layer cannot be determined from a single curvature measurement. However, measuring the change in curvature when the Sn layer is removed by selective etching enables us to obtain $\langle\sigma_{Sn}\rangle h_{Sn}$ for the Sn layer alone.

The whisker density is measured in real time using an optical technique that monitors the scattering of light from the surface [12]. By using oblique illumination in an optical microscope, we can highlight and count surface features over a large area (1 mm × 1 mm) that are too small to resolve directly.

Each set of measurements in Figure 2.3 was made on a series of samples that were grown at the same time under the same conditions to maintain sample uniformity. One

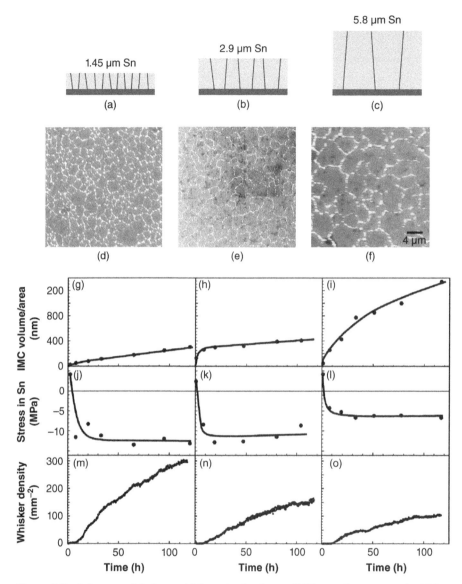

Figure 2.3 Influence of Sn layer thickness/grain size on IMC growth, stress, and whisker density. (a–c) Schematics of sample structure for Sn film thicknesses 1.45 µm, 2.9 µm, 5.8 µm. (d–f) SEM images after the Sn layer has been removed by etching. Bright spots correspond to particles of IMC. Corresponding measurements of (g–i) evolution of IMC volume, (j–l) in-plane stress in Sn layer, and (m–o) whisker density for each layer thickness. Source: Chason et al. [19].

sample was monitored continuously to obtain the whisker density, and other samples were selectively etched after different intervals to measure the stress and IMC volume. Several important features of the whisker formation process can be recognized in these data. The first is that the IMC starts to grow immediately and continues to grow over the entire period of the measurement. The volume increases with a form that becomes parabolic over long times, suggesting that the growth rate becomes limited by diffusion across the IMC layer at the interface [24]. At the same time, as the IMC grows, the stress in the Sn layer becomes increasingly compressive (from an initial tensile stress state). However, even though the IMC continues to grow, the stress saturates at a value of approximately −12 MPa, similar to the yield stress of bulk Sn. Because the IMC continues to grow, this saturation signifies the onset of stress relaxation processes in the Sn layer.

Significantly, whiskers/hillocks start to grow at about the same time when the stress saturates. We observe an initial incubation time during which no whiskers form and then whiskers start to nucleate when the stress reaches its final compressive value. It may appear from this set of measurements that the stress saturation is due to the onset of whisker growth. Indeed, if whisker formation is a stress-relieving mechanism, then extrusion of Sn into whiskers/hillocks should reduce the stress in the layer. However, there are other stress-relieving processes besides whisker formation that can also be occurring, for example, dislocation-mediated plasticity and diffusion-mediated creep.

2.2.2 Dependence on Sn Layer Properties

To obtain a better understanding of how the IMC growth, stress evolution, and whisker growth are related to each other, we look at how these measurements change in samples grown with different Sn layer properties (thickness, grain size, composition, microstructure, etc.). In so doing, we can see how changing one of these quantities, such as the IMC growth rate, affects another one such as the stress.

In the case shown in Figure 2.3, we performed measurements on samples with different nominal thicknesses (1.45, 2.9, and 5.8 μm) [19]. However, it is evident in the SEM images of Figure 2.3d–f that the grain size is not the same for each layer but rather increases with the film thickness (1.42, 1.92, and 3.22 μm, respectively). To separate the effects of grain size and film thickness, we also prepared a series of samples that had different thicknesses but the same grain size. We did this by growing a series of samples to the same thickness (8 μm) and then using an Sn etch to reduce the sample to the desired thickness. During etching, the grain size does not change as it does during deposition. Results of measurements from these samples are shown in Figure 2.4; the grain size in each sample as measured by the SEM images of IMC formation at the interface (Fig. 2.4e–h) remains essentially the same.

The trends in these two sets of data indicate many important features of the IMC growth, stress evolution, and whisker formation. In the case of the IMC growth kinetics, the amount of IMC that forms is relatively weakly dependent on the thickness when the grain size is kept constant (Fig. 2.4i–l). This is consistent with the fact that we believe the IMC grows up from the Sn–Cu interface and depends upon diffusion

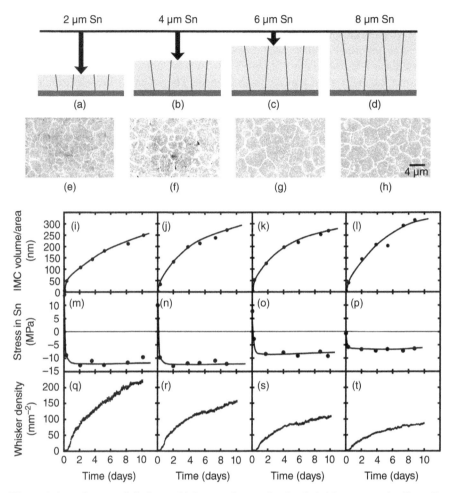

Figure 2.4   Influence of Sn layer thickness when grain size is held constant. An 8 µm Sn film was electroplated on all specimens. The samples were then partially etched to leave Sn layers of difference thickness on the 1 µm vapor deposited Cu layer. The process is shown schematically in (a–d). (e–h) are SEM micrographs of the Sn/Cu interface after selectively etching the Sn layer. Simultaneous measurement of (i–l) IMC volume, (m–p) Sn stress, and (q–t) whisker density for different Sn thicknesses. Source: Chason et al. [19].

across that interface. Therefore, increasing the thickness of the Sn layer would not be expected to affect the IMC growth rate. However, the growth rate does increase significantly when the grain size increases (Fig. 2.3g–i), which suggests that diffusion of Cu through the center of the Sn grains (intragranular region) may be important in controlling the growth rate since IMC forms more slowly there than at the grain boundaries.

The correlation between the IMC growth and the stress is even more revealing about the processes of stress generation and relaxation in the Sn layer. In Figure 2.3, we see that the saturation stress is less for the sample with larger thickness/grain size (Fig. 2.3i) than the thinner sample (Fig. 2.3g) even though the amount of IMC formation is greater (Fig. 2.3f and d, respectively). Since more IMC volume would be expected to generate larger volumetric strain, this suggests that there is more stress relaxation in the sample with larger thickness/grain size than in the thinner sample. For the case of different thicknesses with the same grain size (Fig. 2.4), we see that the saturation stress is lower in the thick sample (Fig. 2.4p) than in the thin sample, even though the amount of IMC is roughly the same. This suggests that the stress relaxation can be enhanced by an increase in the vertical dimension (thickness) as well as the lateral dimension (grain size). In other work on Pb–Sn layers [12], we find that a similar amount of IMC grows as for pure Sn layers of the same thickness. However, the corresponding stress in the layer is very low, indicating that the presence of Pb greatly enhances stress relaxation in Pb–Sn relative to pure Sn.

The fact that the stress in pure Sn layers appears to saturate even though the IMC continues to grow suggests that there is a critical maximum stress determined by the relaxation processes, such as a yield stress in dislocation-mediated plasticity. Indeed, a decrease in yield stress for larger grain size is found in many thin-film systems, such as in the Hall–Petch effect [25]. If, on the other hand, the stress relaxation were due primarily to the whisker formation, it would not be expected that the stress would saturate at a constant value or that this value would depend on the Sn microstructure.

In all our measurements, we observe that lower values of the saturation stress correspond to lower whisker density. This is true for both different grain sizes and different thicknesses. Most importantly, we find it to be true for the samples with different thicknesses that have been etched back from the same initial thickness (Fig. 2.4) [19]. In these samples, because they were etched back from the same original structure, the samples are identical in every way except for thickness. They have the same grain size, texture, and volume of IMC formed. Yet, the saturation stress is different, presumably because the stress relaxation processes is different for each layer thickness.

The correspondence between the stress and the whisker density indicates that whiskering is not the primary stress-relieving mechanism in these Sn layers. If whiskers were the main source of stress relaxation, then we would expect there to be more stress relaxation when there were more whiskers. However, the data in Figure 2.4 indicates the opposite. The films with the most whiskers have the largest compressive stress and the films with the fewest whiskers have the lowest stress. Since the amount of IMC growth is approximately the same in each case, this suggests that the stress is being relaxed by other mechanisms besides whisker growth. In other words, the whiskers appear to be the result of the stress in the layer, but other relaxation processes determine how much stress there will be in the layer.

2.2.3 Stress Relaxation Processes in the Sn Layer

The measurements described earlier point to a picture in which the IMC generates stress, which is then relaxed by processes that depend on the Sn layer structure. To get a better understanding of these processes, it is useful to separate the stress evolution from the IMC formation. We do this by measuring the relaxation behavior of layers

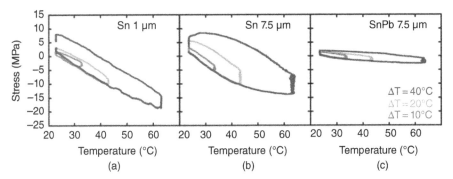

Figure 2.5   Mechanical response of Sn and Pb–Sn films on Si substrates as they undergo multiple heating cycles. The measurements are of stress versus temperature for (a) 1 μm Sn layer (b) 7.5 μm Sn layer, and (c) 7.5 μm SnPb layer.

of Sn deposited directly onto Si substrates with no Cu layer to form IMC. Stress is generated by heating the samples, which leads to strain due to the thermal expansion mismatch between the Sn and Si.

Monitoring the stress evolution during heating cycles enables us to measure the stress relaxation behavior for different Sn layer structures directly, without any IMC formation. A plot of stress versus temperature (which is proportional to strain) is shown in Figure 2.5 for layers of Sn with thicknesses of 1 and 7.5 μm and Pb–Sn with thickness of 7.5 μm [26]. Note that the stress in the 7.5 μm Sn layer saturates at a lower stress value than the 1-μm-thick layer, which is consistent with the enhancement in relaxation inferred from the stress evolution due to IMC formation. The measurements of stress evolution in the Pb–Sn layer indicate substantially more stress relaxation than in the Sn layers, again consistent with the behavior seen with the IMC-induced stress. This enhancement in stress relaxation has been attributed [15] to the different microstructure of Pb–Sn (equiaxed) relative to pure Sn (columnar), which enables relaxation by creep, that is, diffusion of atoms to horizontal grain boundaries.

In pure Sn layers, measurements of the stress versus time while heating indicate that the stress relaxation process follows a power-law creep behavior [23]. However, because of the rapid diffusion of atoms in Sn, we might expect that diffusional creep mechanisms would also play a role. This brings up an interesting question. If grain boundary diffusion is rapid, then why does not the stress relax due to diffusion of atoms to the surface (Coble creep)? There must be some mechanism suppressing the diffusion of atoms to the Sn surface, which is presumably the surface oxide. As Tu has pointed out, whiskers only form on materials that have a strong oxide preventing surface relaxation [24]. We confirmed the role of the surface oxide in preventing stress relaxation by using curvature to measure the stress in the Sn layer with and without the oxide present. In the first case, we used a sputter gun to remove the oxide from the surface of the Sn layer on Cu [7] and observed that the compressive stress in the layer was relieved when the oxide was removed. In other experiments [25], we observed that the compressive stress disappeared when the oxide was removed by chemical etching.

The absence of surface relaxation does not mean that diffusion does not play a role in stress evolution. The fact that whiskers can grow to lengths much greater than the thickness of the layers means that Sn must be transported to the base of the whisker over long distances. The most likely mechanism is the rapid transport of atoms along the network of Sn grain boundaries running through the film. Unfortunately, wafer curvature measurements by nature only measure the average stress in the layers and do not directly provide any information about stress gradients, the driving force for creep. X-ray diffraction measurements [27–29] have shown that there can be stress gradients around the base of whiskers and Tu has proposed a creep-based model of whisker growth [29].

To show that stress gradients can lead to long-range transport of Sn atoms, we designed an experiment to show that whiskers can form far from where the IMC is growing. We did this by depositing Cu over a layer of Sn, but only covering half of the surface [30] (see schematic in Fig. 2.6). The rest of the Sn layer was left uncovered. We found that over time, whiskers started to form on the uncovered Sn surface, away from the Cu layer. SEM/EDAX measurements confirmed that there was no Cu or IMC in the region where the whiskers were forming. Rather, IMC that formed under the Cu layer created stress in the Sn that drove diffusion of Sn atoms into the uncovered Sn layer. This then created stress that leads to the formation of whiskers away from where the IMC was forming. Notably, the distance of the furthest whiskers from the Cu layer increased with time at a rate that was consistent with the grain boundary diffusivity of Sn.

2.2.4 Model of Stress Evolution and FEA Simulation

Based on the experiments described earlier, we have developed the following picture of how stress evolves in the Sn layer, as shown by the schematic in Figure 2.7. The growth of IMC at the Sn–Cu interface creates strain in the Sn layer due to the increase in volume of IMC relative to the native Sn (Fig. 2.7a). The corresponding localized stress is relaxed (Fig. 2.7b) by both dislocation-mediated plasticity in the Sn and diffusional creep, that is, transport of atoms away from the stressed region due to stress gradients. The presence of a native oxide prevents relaxation by diffusion of atoms to the surface.

To see the details of how the stress evolves, these mechanisms have been put into an FEA simulation [31] that includes processes of elasticity, ideal plasticity, grain boundary diffusion, and a capping layer on the surface. The grain structure consists of a hexagonal array of columnar grains. The IMC growth is simulated by the expansion of hemispherical particles at the grain boundary with a rate determined by the measured IMC growth rate.

An image of the stress profile from the FEA simulation is shown in Figure 2.8 [32]. The colors correspond to the biaxial stress in one of the cells of the simulation. Note how the stress is fairly uniform throughout the thickness of the Sn layer. This is due to the diffusion of Sn atoms, which suppresses the creation of large stress gradients and

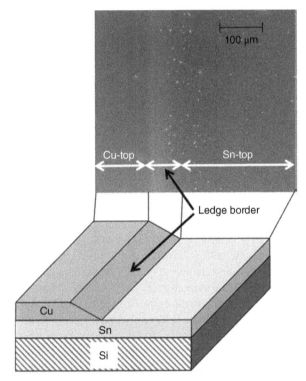

Figure 2.6 A schematic illustration of the ledge sample showing regions of uncoated Sn (Sn-top), Sn covered by Cu (Cu-top), and the transition region (ledge border). A low magnification SEM image of the surface shows that whiskers grow where there is no Cu, indicating the long-range diffusion of Sn from the Cu-top region where IMC forms. Source: Reinbold et al. [30].

leads to a relatively uniform stress profile. The accompanying plot of stress versus depth also shows that the biaxial stress is relatively uniform through the thickness of the layer.

The evolution of the average stress in the simulation as a function of IMC thickness is shown in Figure 2.9. As in the experiments (inverted triangles), the stress saturates at the yield stress of the Sn layer even though the IMC layer continues to grow. In contrast, when the simulation is run without the mechanism of grain boundary diffusion, the plastic zone remains much more localized around the region where the IMC forms. This leads to a much slower saturation of the average stress in the simulations than when grain boundary diffusion is included.

From our measurements and simulations, we believe that the stress saturation is due primarily to the stress relaxation processes of plasticity and diffusional creep. In other words, the stress saturates even without the presence of whiskers. Although whisker formation can be a stress-relieving mechanism, it may not be the primary form of stress relaxation in the layer.

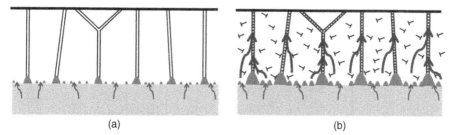

Figure 2.7 (a) Schematic diagram showing Cu diffusing (golden arrows) into Sn layer leading to intermetallic formation at the interface, especially at the Sn grain boundaries. (b) IMC growth induces stress to spread throughout the layer by motion of dislocations and diffusion of Sn atoms (brown arrows) along Sn grain boundaries. Please visit www.wiley.com/go/Kato/TinWhiskerRisks to access the color version of this figure.

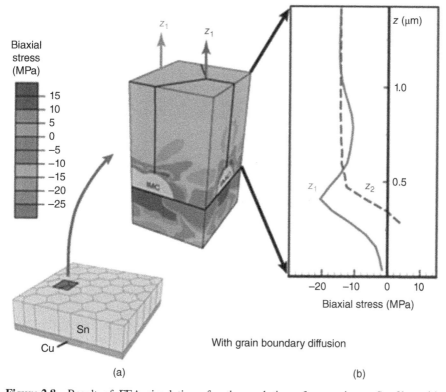

Figure 2.8 Result of FEA simulations for the evolution of stress in an Sn film with columnar microstructure due to IMC growth. The model includes stress relaxation by both elastic–plastic behavior and grain boundary diffusion in the Sn. (a) Contour plots showing the three-dimensional distribution of biaxial stress through the film; the legend relates the color to the stress state. (b) Corresponding stress distributions along the lines z_1 and z_2 indicated on the contour plots. Source: Chason et al. [32]. Please visit www.wiley.com/go/Kato/TinWhiskerRisks to access the color version of this figure.

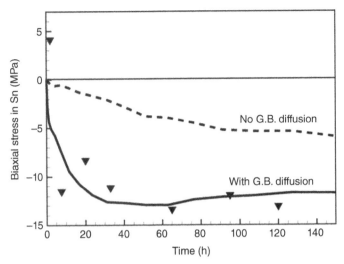

Figure 2.9 Evolution of the average stress in the Sn layer predicted by FEA simulations (lines) and experiment (solid symbols). The dashed line shows FEA predictions for a film that relaxes only by ideal plastic flow in the grain interiors. The solid line shows results for a film that relaxes by both grain boundary diffusion and plastic flow (same as in Fig. 2.13). Source: Reprinted with permission from IEEE Transactions on Electronics Packaging Manufacturing vol. 33, no. 3, pp. 189, July 2010 (© 2010 IEEE) [18].

2.3 RELATION OF STRESS TO WHISKER GROWTH

Although many experiments point to stress as the primary driving force for whisker formation, it does not explain how the stress leads to whisker growth. Tu [29] has described how the reduction in stress at the whisker will lead to gradients that will provide a flux of atoms toward the whisker. But what is the magnitude of the stress gradient? Can it be approximated by assuming that it is uniform and determined by the spacing between whiskers or are there other mechanisms? It also does not address the question of why whiskers form where they do. Are all sites equally likely to form whiskers or is there something special about some that make them more susceptible to whiskering? Answering such questions must be guided by experiments that show how whiskers actually grow. In this section, we present results of whisker growth that include real-time SEM studies of the evolving whisker morphology and discuss what these imply about the mechanisms of whisker formation.

One of the most fundamental questions that must be addressed about whisker formation is why they form at specific sites. One simple possibility is that whiskers form where there is a weakness in the oxide layer since we know that the oxide prevents stress relaxation at the surface. If this is true, then it would suggest that weakening or removing the oxide should lead to whisker formation. In Figure 2.10, we show the effect of removing a section of surface oxide using the FIB. The image in Figure 2.10a was acquired immediately after the oxide was removed (~6 h after it was deposited).

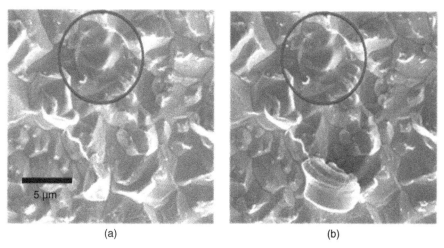

(a)   (b)

Figure 2.10   (a) SEM image of Sn surface with a hole in the oxide layer made by FIB at 6 h after deposition, (b) image after 138 h. No growth is visible where oxide was removed, but hillock is observed to grow approximately 10 μm from the hole.

The circular region highlighted in the figure (diameter 5 μm, depth 10 nm) shows where the FIB modification was performed. The image shown in Figure 2.10b was acquired 138 h later, and it is clear that the surface in and around the region where the oxide was removed has not changed. Similar results were found at multiple sites and with various hole diameters (0.5, 2, and 5 μm). Interestingly, a hillock was observed to form in the unmodified region approximately 10 μm from the hole, indicating that the driving force for hillock growth is still present. We believe that this result clearly indicates that removal of the surface oxide is not sufficient to make whiskers grow. Although removal of large areas of oxide does indeed lead to stress relaxation (as discussed earlier), localized removal does not necessarily lead to whisker or hillock formation.

To understand how whiskers actually grow, we modified an SEM to make multiple images of several regions for an extended period of 4 days. Each region was remeasured at time intervals of 10 min, and the resulting images were assembled into a stop-action movie of the surface evolution. By monitoring five areas of 215 μm × 185 μm, which were 1 mm apart, we were able to capture the real-time growth of both whiskers and hillocks; these videos can be seen at [33].

Several frames from the growth of a whisker are shown in Figure 2.11. The whisker does not start to grow until about 14 h after the film was first deposited, consistent with the incubation time seen in our optical measurements of whisker density (in Figs. 2.3 and 2.4). The whisker clearly grows by the addition of material to its base (not its tip) as has been pointed out by others previously [34, 35]. One important aspect of the growth is that there are no apparent anomalies around the site where the whisker forms before it starts to grow. Nor are there any changes immediately before the whisker starts protruding. In the first few frames of whisker growth, we see a crack in the

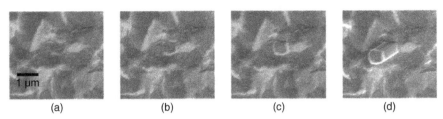

(a)   (b)   (c)   (d)

Figure 2.11   Time series SEM images showing whisker nucleation and growth. Time after deposition: (a) 14 h, (b) 14 h 20 min, (c) 14 h 40 min, (d) 17 h.

oxide develop and spread rapidly around the surface of the grain as the whisker starts to protrude. The surrounding surface remains unchanged as the whisker continues to grow, that is, there is no subsidence of the surface due to transfer of material into the whisker. This confirms that the material that flows into the whisker is not just coming from the local surroundings but diffusing over large distances.

These movies allow us to quantify the length and growth rate of the whisker during its development (Fig. 2.12). The plot of whisker growth rate versus time shows the long incubation period and then rapid growth at the onset of whiskering. After a short period, the growth rate settles down to a steady-state rate that does not change for the duration of the measurement. In the high-vacuum environment that was used for these measurements ($\sim 4.5 \times 10^{-8}$ Torr), we found that features in the form of whiskers (i.e., long thin filaments with no apparent grain growth at their base) grew steadily without changing their rate or direction of growth. This may suggest that the kinks (i.e., change in direction) that are observed in whiskers grown in air may be the result of oxide formation on the surface of the growing whisker. In comparison, the hillock-like features discussed next are much less regular in their growth morphology and growth rate.

The regularity of whisker growth and their growth from specific surface grains suggests that there is something special about whisker-forming grains that enable them to be a sink for the addition of atoms to their base. Two such possible mechanisms are shown in Figure 2.13. The grain that forms whiskers may have a low yield stress so that it pushes out of the film by a process that resembles extrusion [31]. Alternatively, Smetana [36] has suggested that horizontal boundaries at the root of the whisker can act as sinks where atoms can attach. Pushing the whisker upward and out of the films allows the whisker growth to reduce the volume of Sn per unit area of the film and therefore reduce the stress. Indeed, horizontal boundaries at the base of whiskers have been observed in many cross sections of whisker morphology [15].

We have incorporated these mechanisms into an FEA simulation to investigate how the stress evolves during the growth of the whiskers. A mechanism for whisker growth is provided by allowing one grain to have a lower yield stress than other grains or by putting a v-shaped grain boundary underneath it. An FEA image of the stress field surrounding a whisker growing from a v-shaped horizontal grain boundary is shown in Figure 2.14 [31, 32]. The calculation involves the same mechanisms of elasticity, plasticity, and grain boundary diffusion for the simulations described earlier. However, in this case, volumetric strain is applied uniformly through the Sn layer

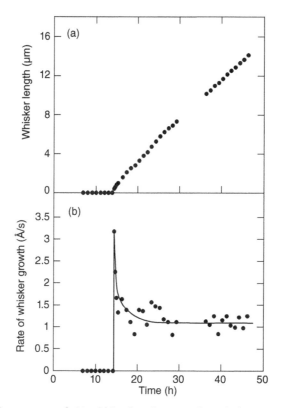

Figure 2.12 Measurement of: (a) whisker length versus time; (b) instantaneous growth rate of whisker versus time. .Source: Jadhav et al. [7]

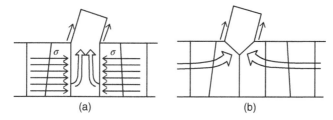

Figure 2.13 Schematic diagram showing two possible mechanisms for whisker growth. (a) Extrusion-based mechanism where the whisker grain has a lower yield stress than the surrounding grains. The vertical broad arrows show the extrusion of Sn atoms via plastic flow. (b) Grain growth-based mechanism in which atoms are added to nonvertical grain boundaries at whisker base. The broad arrows indicate long-range diffusion along the grain boundary network that transports material to whisker grain.

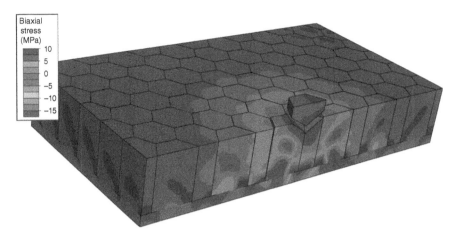

Figure 2.14   FEA model results of stress evolution in SnCu system with whisker growth showing a periodic cell surrounding a single "weak" grain that forms a whisker. The grains in the immediate vicinity of the whiskering grain are relaxed, while the grains further away reach their yield stress (colors correspond to stress levels shown in legend). Source: Chason et al. [32]. Please visit www.wiley.com/go/Kato/TinWhiskerRisks to access the color version of this figure.

at a rate that is correlated with the rate of IMC growth. This generates a biaxial stress in the Sn consistent with that predicted in the previous FEA simulations for growth of discrete IMC particles.

Note that, in this simulation, most of the Sn layer has reached the saturation stress (yield stress of the Sn). The whisker growth only leads to relaxation of the compressive stress in a small region of several grains. However, this is a consequence of the whisker density and the strain rate. Under the simulation conditions shown in the figure, the strain rate is relatively high so that most of the stress in the film is relaxed by plastic deformation in the Sn layer. If the whiskers are closer together or the strain rate is low, compared to the whisker growth rate, then the stress gradient will spread out much further from the whisker base and the formation of whiskers will relieve a significant amount of stress.

An alternative explanation for whisker growth has been suggested by Vianco and Rejent [37]. In their model, whiskers grow due to a process of dynamic recrystallization in which high strain energy density due to plastic deformation leads to the renucleation and growth of strain-free whiskers. This model would suggest that the grain that forms into whiskers must nucleate before the whisker can grow. Although we have not directly observed such nucleation before the formation of the whisker, it is still a potential mechanism for whisker growth.

Not all features on the surface grow in the form of whiskers. We have also observed the morphological evolution of hillocks, as shown in Figure 2.15. The hillock appears to originate from one grain and initially grows primarily in the vertical direction. Unlike the whisker, however, the surface of the hillock rotates by as much as 90° as it grows outward (compare the orientation of the tip of the whisker in Figure 2.15d–a).

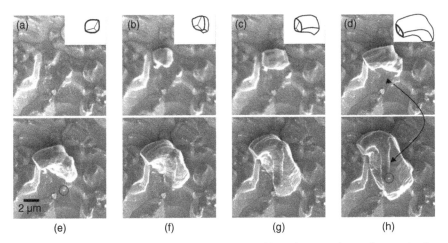

Figure 2.15 SEM images showing hillock growth with surface rotation and extensive lateral grain growth. Time after deposition: (a) 6 h, (b) 12 h, (c) 18 h, (d) 32 h, (e) 44 h, (f) 56 h, (g) 76 h, (h) 138 h. Insets shows schematic of shape evolution highlighting rotation of the original surface. Arrows point to grain boundary features in (d) that are visible as ridges on side of the hillock in (h).

This rotation means that there is a significant amount of plastic deformation occurring in the grain as it grows. It also means that measuring the orientation of the hillock grain after it grows is not an indication of the orientation of the grain before growth. Equally important, in addition to the vertical movement, the base of the hillock grows laterally into the surrounding grains as the volume increases. The vertical and lateral growth appears to alternate, leaving ledges or striations on the hillock surface. This alternation of vertical and horizontal growth has been referred to as a "wedding-cake" morphology by others [17].

2.4 CONCLUSIONS

We have described results of studies designed to identify the fundamental driving forces and mechanisms behind whisker and hillock growth. Simultaneous measurements of IMC growth, stress evolution, and whisker formation show how IMC growth induces stress in the Sn layer and strongly suggests that stress is the primary driving force for whisker growth. Studies of different Sn structures show that the magnitude of stress is controlled by stress relaxation processes in the Sn layer structure, for example, plasticity and diffusion-mediated creep. Enhancing the stress relaxation decreases the magnitude of the steady-state stress and the density of whiskers/hillocks. The surface oxide plays a critical role by suppressing diffusion to the surface as a mechanism for stress relaxation. FEA simulations including processes of plasticity and grain boundary diffusion show how the stress is spread relatively uniformly across the Sn layer even though the IMC growth is localized near the Sn–Cu interface.

SEM movies of whisker growth show that whiskers initiate without any prior weakness or defects apparent in the region around the whisker. Local removal of the oxide by sputtering with the FIB does not cause whiskers to grow, indicating that a weak oxide layer is not sufficient to make whiskers form. The underlying grain structure of the Sn layer is more important. An FEA model was used to model the growth of whiskers from specific sites that can serve as sinks for an atom flux driven by stress gradients, that is, horizontal grain boundaries or grains with low yield stress. The simulations show that the stress gradient depends on the kinetics of the IMC-induced strain and the density of whiskers. We also observed the growth of hillocks that have extensive lateral grain growth coupled with the vertical extrusion. What makes some grains grow in the form of whiskers and others into hillocks is still not understood. Clearly, there is a competition between the extrusion of materials outward and grain growth within the layer. Relating the surface evolution to the underlying microstructure is one of the outstanding challenges in predicting whisker evolution.

ACKNOWLEDGMENTS

The authors gratefully acknowledge the support of the Brown MRSEC (DMR0079964) and NSF (DMR0856229 and DMR1206138) and would like to thank Fei Pei, Jae Wook Shin, Wai Lun Chan, Eric Buchovecky, Lucine Reinbold, Sharvan Kumar, Allan F. Bower, L. B. Freund, Carol Handwerker, and Gordon Barr for their helpful contributions.

REFERENCES

1. Fisher RM, Darken LS, Carroll KG. Accelerated growth of tin whiskers. Acta Metall 1954;2:368–369, 371–373.
2. Galyon GT. Annotated tin whisker bibliography and anthology. IEEE Trans Electron Packag Manuf 2005;28:94–122.
3. Lin SK, Yorikado Y, Jiang JX, Kim KS, Saganuma K, Chen SW, Tsujimoto M, Yanada I. Mechanical deformation-induced Sn whiskers growth on electroplated films in the advanced flexible electronic packaging. J Mater Res 2007;22:1975–1986.
4. Moriuchi H, Tadokoro Y, Sato M, Furusawa T, Suzuki N. Microstructure of external stress whiskers and mechanical indentation test method. J Electron Mater 2007;36:220–225.
5. Cheng J, Chen S, Vianco PT, Li JCM. Quantitative analysis for hillocks grown from electroplated Sn film. J Appl Phys 2010;107:074902-4.
6. Jadhav N, Chason E. Sn whiskers: causes, mechanisms and mitigation strategies. In: *Lead-Free Solders: Materials Reliability for Electronics*. John Wiley & Sons, Ltd; 2012. p 297–321.
7. Jadhav N, Buchovecky E, Chason E, Bower A. Real-time SEM/FIB studies of whisker growth and surface modification. JOM 2010;62:30–37.
8. Glazunova VK, Kudryavtsev NT. An investigation of the conditions of spontaneous growth of filiform crystals on electrolytic coatings. Zh Prikl Khim 1963;36:543–550.

9. Suganuma K, Baated A, Kim K-S, Hamasaki K, Nemoto N, Nakagawa T, Yamada T. Sn whisker growth during thermal cycling. Acta Mater 2011;59:7255–7267.

10. Tu KN. Interdiffusion and reaction in bimetallic Cu–Sn thin films. Acta Metall 1973;21:347–354.

11. Lee BZ, Lee DN. Spontaneous growth mechanism of tin whiskers. Acta Mater 1998;46:3701–3714.

12. Chason E, Jadhav N, Chan WL, Reinbold L, Kumar KS. Whisker formation in Sn and Pb–Sn coatings: role of intermetallic growth, stress evolution, and plastic deformation processes. Appl Phys Lett 2008;92:171901.

13. Reynolds HL, Osenbach JW, Henshall G, Parker RD, Peng S. Tin whisker test development – temperature and humidity effects. Part I: Experimental design, observations, and data collection. IEEE Trans Electron Packag Manuf 2010;33:1–15.

14. Pitt CH, Henning RG. Pressure-induced growth of metal whiskers. J Appl Phys 1964;35:459–460.

15. Boettinger WJ, Johnson CE, Bendersky LA, Moon KW, Williams ME, Stafford GR. Whisker and hillock formation on Sn, Sn–Cu and Sn–Pb electrodeposits. Acta Mater 2005;53:5033–5050.

16. Sobiech M, Welzel U, Mittemeijer EJ, Hugel W, Seekamp A. Driving force for Sn whisker growth in the system Cu–Sn. Appl Phys Lett 2008;93:3.

17. Pedigo AE, Handwerker CA, Blendell JE, Whiskers, hillocks, and film stress evolution in electroplated Sn and Sn–Cu films. In: Electronic Components and Technology Conference; 2008. 58th ECTC; Lake Buena Vista, Florida, USA: IEEE; 2008. pp 1498–1504.

18. Jadhav N, Buchovecky EJ, Reinbold L, Kumar S, Bower AF, Chason E. Understanding the correlation between intermetallic growth, stress evolution, and Sn whisker nucleation. IEEE Trans Electron Packag Manuf 2010;33:183–192.

19. Chason E, Jadhav N, Pei F. Effect of layer properties on stress evolution, intermetallic volume, and density during tin whisker formation. JOM 2011;63:62–67.

20. Oberndorff P, Dittes M, Petit L, Intermetallic formation in relation to tin whiskers, presented at the IPC/Soldertec International Conference on Lead-Free Electronics Towards Implementation of the ROHS Directive, Brussels, 2003.

21. Zhang W, Egli A, Schwager F, Brown N. Investigation of Sn–Cu intermetallic compounds by AFM: new aspects of the role of intermetallic compounds in whisker formation. IEEE Trans Electron Packag Manuf 2005;28:85–93.

22. Floro J, Chason E, Lee S, Twesten R, Hwang R, Freund L. Real-time stress evolution during $Si_{1-x}Ge_x$ heteroepitaxy: dislocations, islanding, and segregation. J Electron Mater 1997;26:969–979.

23. Freund LB, Suresh S. *Thin Film Materials: Stress, Defect Formation and Surface Evolution*. 1st ed. Cambridge, U.K.: Cambridge University Press; 2009.

24. Reinbold L, Chason E, Jadhav N, Kelly V, Holmes P, Shin JW, Chan WL, Kumar KS, Barr G, Degradation Processes in Nanostructured Materials Book Series: Materials Research Society Symposium Proceedings: Cambridge, U.K.; 2006. pp 197–207.

25. Shin JW, Chason E. Stress behavior of electroplated Sn films during thermal cycling. J Mater Res 2009;24:1522–1528.

26. Jadhav N, Wasserman J, Pei F, Chason E. Stress relaxation in Sn-based films: effects of Pb alloying, grain size, and microstructure. J Electron Mater 2012;41:588–595.

27. Sobiech M, Wohlschlogel M, Welzel U, Mittemeijer EJ, Hugel W, Seekamp A, Liu W, Ice GE. Local, submicron, strain gradients as the cause of Sn whisker growth. Appl Phys Lett 2009;94:221901.

28. Choi WJ, Lee TY, Tu KN, Tamura N, Celestre RS, MacDowell AA, Bong YY, Nguyen L. Tin whiskers studied by synchrotron radiation scanning X-ray micro-diffraction. Acta Mater 2003;51:6253–6261.

29. Tu KN, Chen C, Wu AT. Stress analysis of spontaneous Sn whisker growth. J Mater Sci Mater Electron 2007;18:269–281.

30. Reinbold L, Jadhav N, Chason E, Kumar KS. Relation of Sn whisker formation to intermetallic growth: results from a novel Sn–Cu "bimetal ledge specimen". J Mater Res 2009;24:3583–3589.

31. Buchovecky EJ, Du NN, Bower AF. A model of Sn whisker growth by coupled plastic flow and grain boundary diffusion. Appl Phys Lett 2009;94:3.

32. Chason E, Jadhav N, Pei F, Buchovecky E, Bower A. Growth of whiskers from Sn surfaces: driving forces and growth mechanisms. Prog Surf Sci 2013;88:103–131.

33. Chason E, Jadhav N. (2010). Research activities in Prof. Eric Chason's Laboratory. Available: http://www.engin.brown.edu/faculty/Chason/research/whisker.html.

34. Koonce SE, Arnold SM. Growth of metal whiskers. J Appl Phys 1953;24:365–366.

35. Lindborg U. Observations on growth of whisker crystals from zinc electroplate. Metall Trans A 1975;6:1581–1586.

36. Smetana J. Theory of tin whisker growth: "The end game". IEEE Trans Electron Packag Manuf 2007;30:11–22.

37. Vianco PT, Rejent JA. Dynamic recrystallization (DRX) as the mechanism for Sn whisker development. Part I: A model. J Electron Mater 2009;38:1815–1825.

3

APPROACHES OF MODELING AND SIMULATION OF STRESSES IN Sn FINISHES

PENG SU

Juniper Networks, Sunnyvale, California, USA

MIN DING

Freescale Semiconductor, Austin, Texas, USA

3.1 INTRODUCTION

Many suppliers have selected pure Sn to replace the current SnPb finish for leaded components since it provides good solderability and does not require major changes of the existing manufacturing infrastructure. However, the spontaneous whisker growth on Sn finish surfaces is a serious reliability issue. It is believed that the compressive stress developed in the Sn finish is the major driving force for the whisker growth, which acts as a stress relief mechanism [1–4]. It is also believed that the continuous increase of this stress is due to intermetallic compound formation at the Sn/lead-frame interface [3, 4]. Recently, Barsoum et al. proposed that the compressive stress is due to the oxidation of Sn [1]. The oxygen diffuses into the grain boundaries and forms Sn oxides, which generates the compressive stress in the Sn film by increasing the volume of the confined Sn film.

The other whisker growth scenario is temperature cycling. Whiskers grow from the Sn finish on electronic packages during power cycling. The growth speed is normally much faster than the spontaneous growth during storage. Air-to-air thermal cycling

Mitigating Tin Whisker Risks: Theory and Practice, First Edition.
Edited by Takahiko Kato, Carol A. Handwerker, and Jasbir Bath.
© 2016 John Wiley & Sons, Inc. Published 2016 by John Wiley & Sons, Inc.
Companion website: www.wiley.com/go/Kato/TinWhiskerRisks

test (AATC) has been widely used for evaluating the growth rate of whiskers on lead (Pb)-free finishes including pure tin (Sn) and high-Sn alloys. Recently, it has been adopted by JEDEC as one of the three acceleration tests for whisker growth [5, 6].

The thermal stress due to the mismatch of coefficient of thermal expansion (CTE) between Sn and the lead-frame material is the dominant source of stress in the Sn finish during temperature cycling. To resolve the thermal stress, the mechanical properties of the Sn finish and lead frame need to be known. While the lead frame, most of the time made of Cu, can be considered as an isotropic material, it is unrealistic to make the same assumption for Sn finish. On the grain level, white Sn (β-Sn) has an anisotropic body-centered tetragonal structure with its c-axis much shorter than its a-axis ($c/a = 0.546$). The elasticity anisotropy and thermal expansion anisotropy are significant [7]. On the thin-film level, not all the grains in the Sn finish are aligned to the same orientation and certain texture exists, which can be characterized with X-ray diffraction (XRD). Therefore, the grain orientation information, in addition to the grain size and shape of the Sn finish, should be included when evaluating the whisker growth propensity of Sn finishes.

There has been some work studying the stress in Sn finish analytically [4]. However, the complexity of the problem makes the finite element method (FEM) a better approach for the same purpose, although various assumptions have to be made [1, 2]. In this chapter, the constitutive model of Sn grain will be discussed first as the foundation of FEM modeling. After that, the strain energy density (SED) will be introduced as the failure criterion. A common technique in Sn finish FEM modeling, grain rotation, will also be introduced. Two FEM examples will then be given to explore the impact of the grain orientation on whisker growth in Sn finishes. The first example focuses on the elastic behavior of three immediately adjacent grains without considering the effects of grains in longer distances, while the second example considered elastoplastic behavior of a much larger group of grains.

3.2 CONSTITUTIVE MODEL

3.2.1 Elasticity Anisotropy

β-Sn is a body-centered tetragonal crystal structure, which satisfies the class $4mm$ symmetry. The generalized Hooke's law for a tetragonal single crystal with class $4mm$, $\overline{4}2m$, 422, or $4/mmm$ symmetry can be expressed by [8]

$$
\begin{bmatrix} \sigma_{11} \\ \sigma_{22} \\ \sigma_{33} \\ \sigma_{23} \\ \sigma_{31} \\ \sigma_{12} \end{bmatrix} = \begin{bmatrix} c_{11} & c_{12} & c_{13} & & & \\ c_{12} & c_{11} & c_{13} & & & \\ c_{13} & c_{13} & c_{33} & & & \\ & & & c_{44} & & \\ & & & & c_{44} & \\ & & & & & c_{66} \end{bmatrix} \begin{bmatrix} \varepsilon_{11} \\ \varepsilon_{22} \\ \varepsilon_{33} \\ \varepsilon_{23} \\ \varepsilon_{31} \\ \varepsilon_{12} \end{bmatrix}. \tag{3.1}
$$

The elastic stiffness matrix elements (in gigapascal) of β-Sn near room temperature are $c_{11} = 109.4$, $c_{33} = 107.8$, $c_{12} = 57.67$, $c_{13} = 34.76$, $c_{66} = 26.75$, and

$c_{44} = 2.56$ [7]. The elastic stress–strain relation of an orthotropic material is expressed in Equation 3.2 [9]

$$
\begin{bmatrix}
\varepsilon_{xx} \\
\varepsilon_{yy} \\
\varepsilon_{zz} \\
\gamma_{xy} \\
\gamma_{xz} \\
\gamma_{yz}
\end{bmatrix}
=
\begin{bmatrix}
\frac{1}{E_x} & \frac{-v_{xy}}{E_y} & \frac{-v_{xz}}{E_z} & & & \\
\frac{-v_{yx}}{E_x} & \frac{1}{E_y} & \frac{-v_{yz}}{E_z} & & & \\
\frac{-v_{zx}}{E_x} & \frac{-v_{zy}}{E_y} & \frac{1}{E_z} & & & \\
& & & \frac{1}{G_{xy}} & & \\
& & & & \frac{1}{G_{yz}} & \\
& & & & & \frac{1}{G_{xz}}
\end{bmatrix}
\begin{bmatrix}
\sigma_{xx} \\
\sigma_{yy} \\
\sigma_{zz} \\
\sigma_{xy} \\
\sigma_{xz} \\
\sigma_{yz}
\end{bmatrix}
\tag{3.2}
$$

With the constraints

$$
\frac{v_{yx}}{E_x} = \frac{v_{xy}}{E_y}, \frac{v_{zx}}{E_x} = \frac{v_{xz}}{E_z}, \frac{v_{zy}}{E_y} = \frac{v_{yz}}{E_z}
\tag{3.3}
$$

The elastic compliance matrix S (with elements s_{ij}) is the inverse of C (with elements c_{ij})

$$
S = C^{-1} =
\begin{bmatrix}
s_{11} & s_{12} & s_{13} & & & \\
s_{12} & s_{11} & s_{13} & & & \\
s_{13} & s_{13} & s_{33} & & & \\
& & & s_{44} & & \\
& & & & s_{44} & \\
& & & & & s_{66}
\end{bmatrix}
\tag{3.4}
$$

The conversion from elastic compliance matrix elements to the engineering constants (E_x, G_{xy}, etc.) is the following [10]:

$$
\begin{aligned}
E_x &= E_y = 1/s_{11}, & E_z &= 1/s_{33}, \\
G_{xy} &= G_{yz} = 1/s_{44}, & G_{xz} &= 1/s_{66}, \\
v_{xz} &= v_{yz} = -s_{13}/s_{33}, & v_{xy} &= -s_{12}/s_{11}.
\end{aligned}
\tag{3.5}
$$

Therefore, the elastic behavior of β-Sn single crystal can be described by the engineering constants with $E_x = E_y = 76.20$ GPa, $E_z = 93.33$ GPa, $G_{xy} = 26.75$ GPa, $G_{yz} = G_{xz} = 2.56$ GPa, $G_{xy} = 0.473$, $G_{xz} = 0.170$, and $G_{yz} = 0.208$. The CTE is alpha $\alpha_x =$ alpha $\alpha_y = 15.8 \times 10^{-6}/°C$, and alpha $\alpha_z = 28.4 \times 10^{-6}/°C$ [7].

3.2.1.1 Plasticity Effect

A good treatment of Sn grain plasticity should follow the crystal plasticity theory [11, 12]. However, this theory makes additional assumptions on slip system resistance parameters. An experimental calibration of its constitutive parameters is needed before the crystal plasticity model can be used. Unfortunately, there is no β-Sn crystal plasticity model existing at present.

Isotropic hardening is normally a good approximation of low-melting-point poly-crystalline metals such as Sn [13]. A bilinear isotropic hardening rule can used to approximate the plastic behavior of Sn. This approximation does not consider the crystal orientations and slip systems, which may not be realistic, but still serves as a first-order approximation due to the lack of material property data and a better plastic model. According to Diulin et al. [14], single-crystal β-Sn (with 0.01–0.53 at.% Zn) yields at 9.2 MPa (uniaxial tension) at room temperature, and the tangent modulus is 85 MPa in the bilinear isotropic hardening model. These data were measured on single-crystal Sn samples with proper axis of elongation such that the maximum shear stress was attained in the slip system (100)<010>. These two parameters can be used to account for the plastic behavior of β-Sn on copper lead frames.

3.3   STRAIN ENERGY DENSITY

Although compressive stress is believed to be the driving force of Sn whisker growth [1–4], its tensorial nature makes the stress hard to serve as a one-parameter criterion for whisker growth. However, if energy-based criterion is used, it is possible to use one parameter to describe the risk of whisker growth. Therefore, the SED can be used as the parameter to describe the whisker growth propensity. SED is defined by

$$w = \int \sigma_{ij}\, d\varepsilon_{ij} \tag{3.6}$$

Here, summation of repeated indices is assumed. In this definition, elastic and inelastic SED is included. SED actually combines both stress and strain effect. SED has a unit of joule per cubic meter (=pascal). Megapascal will be used as the SED unit in this work. In some FEM software, such as ANSYS, an element table manipulation is used to generate the SED in the postprocessing stage.

3.4   GRAIN ORIENTATION

As mentioned earlier, not all the grains in the Sn finish have the same orientation. Consequently, each grain in the FEM model needs to be rotated to the proper orientation within the laboratory coordinate system in order to simulate the texture of the Sn finish. Euler angles are normally used to describe the rotations of coordinates [15]. However, the definitions of Euler angles are different by different authors. The FEM software ANSYS defines Euler angles α, β, and γ according to Figure 3.1 [9]. It is consisted of three rotations. The first rotation is about the z-axis in the x–y plane from x-axis toward y-axis to form a new coordinate system (x_1–y_1–z_1) with an angle α. The second rotation is about the x_1-axis in the y_1–z_1 plane from y_1-axis toward z_1-axis to form coordinate system (x_2–y_2–z_2) with an angle β. The third rotation is about the y_2-axis from z_2-axis toward x_2-axis to form (x_3–y_3–z_3) with an angle γ [9].

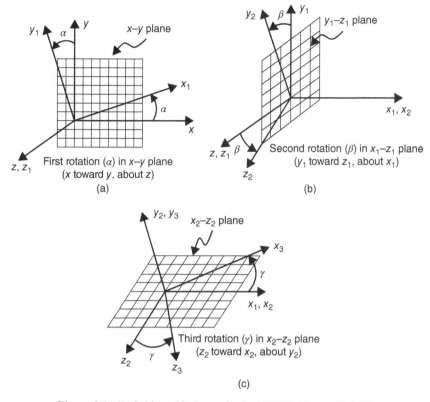

Figure 3.1 Definition of Euler angles in ANSYS. Source: Ref. [9].

The three rotation matrices of the Euler angles are defined by [15]

$$R_1(\alpha) = \begin{bmatrix} \cos(\alpha) & \sin(\alpha) & 0 \\ -\sin(\alpha) & \cos(\alpha) & 0 \\ 0 & 0 & 1 \end{bmatrix} \qquad (3.7)$$

$$R_2(\beta) = \begin{bmatrix} 1 & 0 & 0 \\ 0 & \cos(\beta) & \sin(\beta) \\ 0 & -\sin(\beta) & \cos(\beta) \end{bmatrix} \qquad (3.8)$$

$$R_3(\gamma) = \begin{bmatrix} \cos(\gamma) & 0 & -\sin(\gamma) \\ 0 & 1 & 0 \\ \sin(\gamma) & 0 & \cos(\gamma) \end{bmatrix} \qquad (3.9)$$

By using the given three rotation matrices, one can rotate any vector from one direction to a given direction in a 3D space. To assign the proper orientation to each Sn grain, the following procedure is used in the model. First, it is assumed that every

Sn grain is originally oriented with its (100) on the x-axis, (010) on the y-axis, and (001) on the z-axis of the laboratory coordinates (here, z-axis is the out-of-plane direction of the sample). We assume that the normal of the (hkl) planes is described by the three direction cosines (a, b, c) in the lab coordinate system in this assumed "original" orientation. Then, one can make three rotations according to the Euler angles defined earlier to orient each crystal grain to get the (hkl) texture orientation. This is equivalent to rotate a vector with its direction cosines (a, b, c) to the final direction of the z-axis of the lab coordinates with direction cosines of $(0, 0, 1)$. The rotation process is expressed in terms of the rotation matrices and direction cosines in Equation 3.10:

$$\begin{bmatrix} 0 \\ 0 \\ 1 \end{bmatrix} = R_3(\gamma)R_2(\beta)R_1(\alpha) \begin{bmatrix} a \\ b \\ c \end{bmatrix} \tag{3.10}$$

Equation system 3.10 seems formidable. However, it can be solved after some algebra. Solving it, one has the Euler angles

$$\beta = -\arctan\left(\frac{b\cos\alpha - a\sin\alpha}{c}\right) \tag{3.11}$$

$$\gamma = \arctan\left(\frac{a\cos\alpha + b\sin\alpha}{\sqrt{(b\cos\alpha - a\sin\alpha)^2 + c^2}}\right) \tag{3.12}$$

where the first Euler angle α is arbitrary. It is due to an arbitrary rotation of the grain around the (hkl) axis.

In the FEM model of ANSYS, the grain orientation (hkl) is assigned to each element by rotating its element coordinate system according to the Euler angles α, β, and γ determined by Equations 3.11 and 3.12.

3.5   FINITE ELEMENT MODELING OF TRIPLE-GRAIN JUNCTION

3.5.1   Microstructural Observation

It has been observed commonly that immediately adjacent to or near every whisker grain, the top surface of the original plated Sn grains often becomes much smoother and is recessed from the original position. These features suggest that these grains have been consumed by the nearby whisker grains, and the diffusion paths of the Sn atoms are most likely the grain surface and the grain boundaries. One of such an example is shown in Figure 3.2. The original grain in the center of the image has almost been completely consumed by the whisker grains next to it, forming a very deep recess.

The damage process of the originally plated grains is likely faster at grain boundaries and multigrain junctions. At these locations, the concentration of stresses could occur, and Sn atoms can diffuse away very quickly as they are already on one of the faster diffusion paths–grain boundaries. In Figure 3.3, the grain on the left is showing

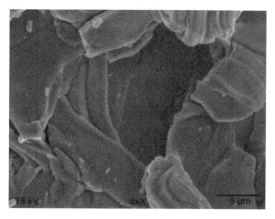

Figure 3.2   A plated Sn grain that has been consumed by neighboring whisker grains.

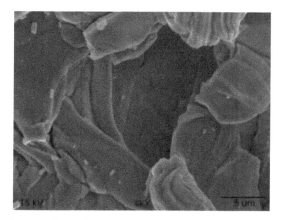

Figure 3.3   The damage of two adjacent as-plated grains, with concentration points at grain boundary or multigrain junctions. A whisker is growing at the upper left corner.

a recess on the grain boundary, while the one on the right is showing damage at two multigrain intersections. Sn atoms originally at these locations apparently have diffused away through the grain boundaries or grain surface toward the whisker grains, one of which is near the grain on the left.

It has been demonstrated that the grain orientation mixes of the as-plated Sn finishes have a strong effect on whisker density. As whisker growth is a stress-induced phenomenon, the difference in whisker density suggests that the stress levels within these samples must be different. The impact of grain orientation on stress concentration levels has been studied extensively for materials with highly anisotropic mechanical properties [16–19]. At multigrain junctions, the grains can be stressed very differently if the rotation configurations of the grains are different [16]. In the case of Sn finish, during the same temperature cycle, stress levels could be much higher at certain grain junctions because of the anisotropic E and CTE. As a result, damage (yield/slip) of the grains would occur, providing the necessary conditions for the nucleation of a whisker grain. Additionally, the fact that the consumption of the as-plated grains is highly localized also indicates that the effects of grain orientation do not propagate too far away from the stress concentration points. In Figure 3.3, for example, the grains at the lower left corner of the image show little signs of damage, although they are immediately adjacent to the two damaged grains in the center.

These observations suggest that for the purpose of predicting the possibility of whisker nucleation, one only needs to focus on the behavior of a small group of immediately adjacent grains, without considering the effects of grains in longer distances.

3.5.2   Finite Element Modeling

Finite element software that accepts anisotropic material properties is necessary to simulate Sn finish, such as ABAQUS and ANSYS. The finite element model examples in this chapter used the commercial software ANSYS. ANSYS provides an anisotropic element called Solid64, which can account for the elastic anisotropy by taking the full elastic stiffness matrix as its input. However, this element does not support plasticity. Since the tetragonal structure of β-Sn falls in the category of orthotropic materials, one can take advantage of the regular continuous solid elements of ANSYS such as Solid45 and Solid185. The 3D eight-node solid element Solid185 is used here. This element has large deformation capability and supports orthotropic elasticity and plasticity. It is also proved that the convergence performance is better than other elements, such as Solid45.

In this FEM modeling work, SED is used as the major parameter for comparison. It is assumed that once the SED level reaches a threshold level SED_{th} within a grain, damage would start to occur and a whisker would nucleate next to this grain. Since the SED level in each grain is mostly determined by its immediately adjacent grains, we will only need to calculate the SED levels in small groups of grains.

One example of this methodology is shown in Figure 3.4. The top surface of the Sn finish has been cleaned with a focused ion beam (FIB) so that the grain boundaries can be clearly seen. All of the multigrain junctions have been circled for clarity. In order to assess the probability of whisker growth at each of the grain junctions, one can treat each of the circled grain junctions as discrete systems and calculate the SED levels in that system individually. If the SED level in any of the grains exceeds SED_{th}, we can then assume a whisker would nucleate at that junction. After the SED is calculated for each of the grain junctions, prediction of the whisker density on the Sn finish within this area is possible.

To simplify the model, the grain junctions are assumed to be triple-grain junctions. The geometry of the finite element model consists of a 10-µm-thick layer of Sn and a 50-µm-thick Cu lead frame. The top surface of the finite element model is shown in Figure 3.5. The bottom 50 µm (away from the reader) is modeled as pure Cu with isotropic elasticity. The top 10 µm (facing the reader) is Sn with anisotropic elasticity.

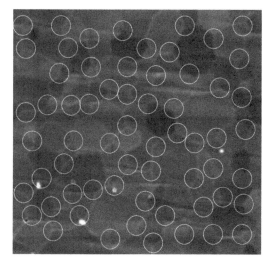

Figure 3.4   SEM image of the top surface of Sn finish after FIB.

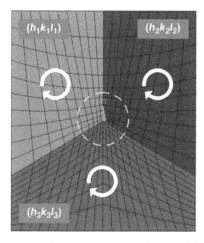

Figure 3.5   The three-grain structure simulated in the finite element model.

Grain structure is implemented by rotating the local stiffness matrix for each of the three differently colored blocks.

Each of the grains is rotated 360° in-plane with 30° steps. Therefore, a total of $12^3 = 1728$ calculations are performed. For each rotation configuration, only the cooling step of the AATC test (from 85 to $-55°C$) is calculated. After the thermal load is applied, the SED level within each grain, as well as the SED near the grain junction (within the circle in the dashed line), is calculated for each of the grains. The areas within this circle are important because it is the most likely nucleation site for whisker grains.

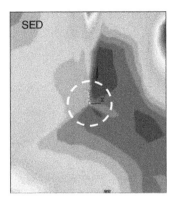

Figure 3.6   An example of the SED distribution within a triple-grain system.

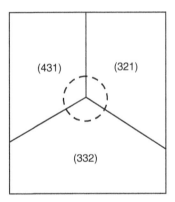

Figure 3.7   The relative relationship of the three grains as simulated in the model.

One example of the SED distribution calculated by the finite element model is plotted in Figure 3.6. The average SED of each of the grains, as well as the SED levels in the circled areas, can be separately extracted from the model. In this plot, the different levels of SED across the grain boundaries can be clearly seen.

3.5.3   Effects of Grain Rotation on SED Distribution

As an example, the results from one of the triple-grain combinations, (431)–(321)–(332), is presented. To be brief, (332) has a fixed rotation angle, and only the SED distributions in grain (431) are plotted. Figure 3.7 shows the configuration of the three grains. Both of the (431) and (321) grains are rotated in 30° steps for 12 times, and the SED levels are individually calculated for each of the 144 configurations.

Figure 3.8a is the mean SED in grain (431) as affected by the rotation of grain (321). For each of the 30° rotation of the (321) grain, there are 12 SED values because

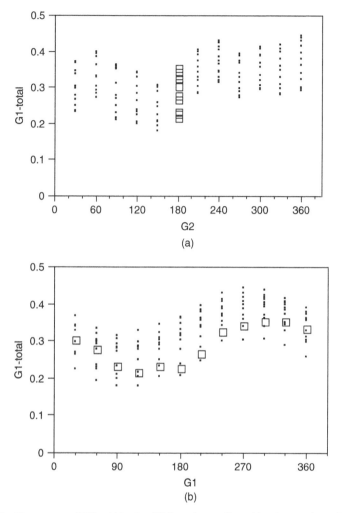

Figure 3.8   The average SED within the (431) grain as affected by the rotation of (431) and the neighboring (321) grain. The third grain (332) is fixed with no rotation. (a) Effect of the rotation of (321) on the SED distribution within (431). At each angle, the (431) grain is rotated in 30° steps for 12 times. (b) Effect of the rotation of (431) on the SED distribution within (431). At each angle, the neighboring (321) grain is rotated in 30° step for 12 times.

the (431) grain is rotated 12 times between 0° and 360°. The general trend of the grain average SED is that the mean value of each of the 12-point data set is somewhat modulated by the rotation of the (321) grain, and within each of the groups, the range of the SED remains relatively consistent.

The same set of data is plotted differently in Figure 3.8b, with the rotation angle of the (431) grain as the X-axis. The modulating effect of the (431) rotation on the SED distribution is much more apparent. The data points marked with the squares are the same for both Figure 3.8a and b. The variation of these data is clearly seen in Figure 3.8b.

The SED levels at the center of the triple junction (within the circles in Figure 3.4) are also calculated. SED levels in this region are important because a high level of SED will induce damage (slip or yield) in the adjacent grains, which will consequently assist the nucleation of whisker grains at this site. Figure 3.9a and b are the

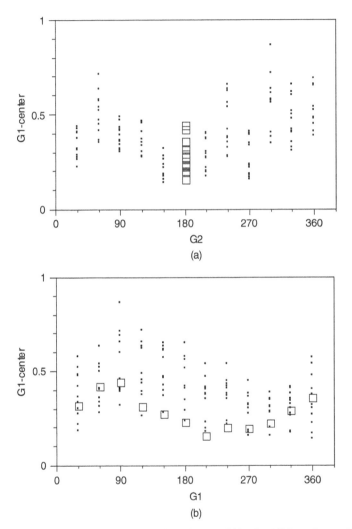

Figure 3.9 The SED distribution of the circled area within the (431) grain as affected by the rotation of (431) and the neighboring (321) grain. The third grain (332) is fixed with no rotation. (a) Effect of the rotation of (321) on the SED distribution at the corner of the (431) grain. At each angle, the (431) grain is rotated in 30° step for 12 times. (b) Effect of the rotation of (431) on the SED distribution within the circled area of the (431) grain. At each angle, the neighboring (321) grain is rotated in 30° steps for 12 times.

mean SED of the circled portion of the (431) grains as affected by the rotation of the (321) and (431) grains. Again, the rotation of the (431) grain itself has a much more apparent modulating effect on the level of the SED in the (431) grain. As shown in Figure 3.8, one group of data is plotted with the square symbol to show the correspondence of data sets.

3.5.4 Construction of a Statistics-Based Predictive Model

The same calculation process for the (431)–(332)–(321) system can be repeated for each possible triple-grain combination for Sn finishes with a given grain orientation mix. Assume that the SED levels in each grain dictate the probability for a whisker to nucleate near that grain, then once the entire set of SED distribution data is obtained, one can compare them with a threshold value SED_{th} to assess the overall whisker growth probability for each of the three-grain systems. For the (431)–(332)–(321) system, for example, if 80% of the $12^3 = 1728$ rotations have SED above SED_{th}, one can say that the probability for whisker growth near this combination is 80%. The summation of all the possible triple-grain combinations in a Sn finish will yield a probability description of its whisker propensity.

3.6 FINITE ELEMENT MODELING OF Sn FINISH WITH MULTIPLE GRAINS

3.6.1 3D Modeling of Sn Finish

In this example, a dimensional (3D) FEM model is established to simulate columnar Sn grains on a Cu lead frame. The grain structure is explicitly established in the model by using the Voronoi diagram method, which will be explained. Sn elasticity on each grain is modeled by the elastic stiffness matrix of Sn single crystal. The crystal orientations are assigned to the grains according to the information collected by XRD. The stress along with the SED distributions in grains is calculated by the FEM model for different configurations of Sn microstructure. The effect of the Sn plasticity will be explored using a simple bilinear isotropic hardening stress–strain relation. The creep effect will not be considered in the current formulation. This work focuses on the temperature cycling test of Sn whisker growth. Although creep is significant for Sn at high temperature, for moderate temperature ranges (-55 to $85°C$) and normal loading speed (a few cycles per hour) in thermal cycling tests, the general picture is qualitatively correct without considering creep.

3.6.2 Geometry Modeling

In order to describe the geometric characteristics of the textural Sn finish on a Cu lead frame, a Voronoi diagram method is used to generate the grain pattern. Voronoi diagrams are recently attracting the attention of material scientists [20–22]. A Voronoi

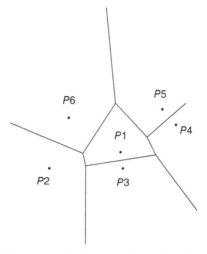

Figure 3.10   A Voronoi diagram with six generating points. Generating point P1 is in a Voronoi cell.

tessellation can be generated in the following way. Suppose we have a finite number of distinct points, which are called the generating points, in a given space. For each generating point, draw perpendicular lines to bisect the lines joining neighboring generating points. The bisecting lines form cell walls enclosing that generating point. An example is shown in Figure 3.10. For generating point P1, five perpendicular lines are drawn to bisect the lines (not shown in the figure) connecting the neighboring points (P2, P3, P4, P5, and P6). Point P1 is enclosed in a cell with the bisecting lines as the cell walls. The cell is called the Voronoi region or Voronoi cell. The Voronoi cells are convex polygons. This process goes on for all other points until every point is enclosed in a Voronoi region, except the edge points, which are only separated by ray lines. In Figure 3.10, all points are edge points except P1.

The Sn grains in this work are modeled as a columnar structure, which can be extruded from a 2D Voronoi diagram. The Voronoi diagram is generated using the mathematics software Mathematica [23]. The coordinates of the generating points are generated by the random number generator of Mathematica. The Voronoi diagrams are generated by calling its Computational Geometry add-on package. The average grain size is controlled by putting a proper number of generating points in a known area. The Mathematica program generates the coordinates of the vertexes of the Voronoi diagram and the connection relationship of the vertexes. To generate an FEM mesh in the finite element software ANSYS [9], an ANSYS Parametric Design Language [9] script is used to input the coordinates and connection relationship of the Voronoi diagram.

Figure 3.11 shows a typical grain pattern with 50 grains generated by using the Voronoi tessellation algorithm. A scanning electron microscope (SEM) picture is also shown in the figure as a computer grain pattern is fairly similar to the grain pattern in the SEM picture.

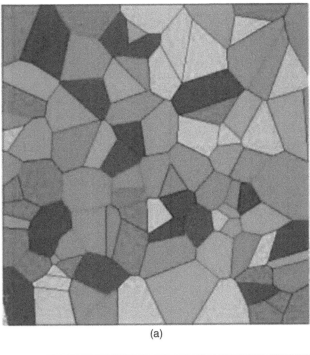

(a)

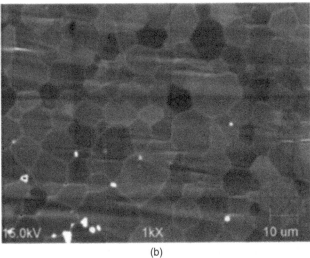

(b)

Figure 3.11 (a) A typical grain pattern (Voronoi diagram) with 50 grains generated by computer using Mathematica and ANSYS. The color is arbitrarily assigned to enhance the visibility. (b) An SEM image of actual grains of the Sn finish plated on a copper lead frame. Please visit www.wiley.com/go/Kato/TinWhiskerRisks to access the color version of this figure.

TABLE 3.1   X-Ray Diffraction Intensities and Texture Fractions for Two Samples from Different Plating Baths

(hkl)	I_0(hkl)	2θ (°)	I(hkl) (Bath-A)	T_f(hkl) (Bath-A)	I(hkl) (Bath-B)	T_f(hkl) (Bath-B)
(101)	90	32.02	5.30	0.01	0.40	0
(220)	34	43.87	19.10	0.07	100.0	0.43
(211)	74	44.90	58.00	0.10	66.40	0.13
(112)	23	62.54	100.0	0.58	0	0
(321)	20	64.58	35.50	0.24	45.60	0.33
(420)	15	72.41	0	0	8.90	0.09
(411)	15	73.20	0	0	1.30	0.01
(312)	20	79.47	0	0	0.90	0.01

3.6.3   Texture and Grain Orientation

The texture of the Sn coating is modeled by assigning a specific grain orientation to each Voronoi cell. The orientation assignment is based on the XRD data of actual samples. The texture fraction T_f(hkl) for an (hkl) orientation is calculated by

$$T_f(hkl) = \frac{I(hkl)/I_0(hkl)}{\sum_{hkl}[I(hkl)/I_0(hkl)]} \tag{3.13}$$

where I(hkl) and I_0(hkl) are the intensities of (hkl) reflections measured on the textural sample and a standard β-Sn powder sample, respectively. Table 3.1 lists the texture fractions of Sn finishes deposited using two different plating baths. In that table, the 2θ is the diffraction angle. The intensities of the peaks are in arbitrary units with the maximum peak normalized to 100 for each sample itself. For powder sample, the maximum intensity peak is not listed there. The XRD results clearly show that the textures are different in these two samples.

Using the measured texture fraction T_f(hkl), the previously generated cells (representing grains) of the Voronoi diagram are assigned corresponding grain orientations. The XRD measurement T_f(hkl) value is an integration value without specific grain information. It is not as detailed as other methods, such as electron backscattering diffraction, which can index each grain. However, XRD is a fast measurement technique and can be incorporated in the manufacturing process. The XRD data also provide great flexibility for grain assignment in the FEM modeling. It is only required to maintain the texture fraction value of the modeled structure conforming to the measured T_f(hkl) values.

3.6.4   Finite Element Model

The Mathematica-generated Voronoi diagram grain pattern is read into ANSYS. A columnar Sn grain structure is assumed. An extrusion of the Voronoi diagram

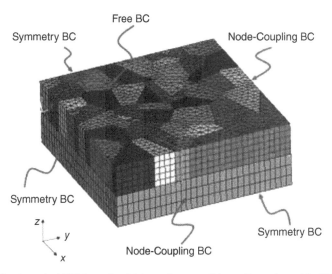

Figure 3.12 A typical FEM mesh with boundary conditions. Sn grains with different grain orientations are color-coded. The Cu lead frame is modeled as an isotropic material. Please visit www.wiley.com/go/Kato/TinWhiskerRisks to access the color version of this figure.

forms a three-dimensional mesh. For thin Sn films, say about 10 μm, columnar grain pattern is a very good assumption, although at some portion, there are occasionally some stacking grains in the film thickness direction. Sn is modeled with the elasto-plastic model described earlier. Cu lead frame is modeled as a linear elastic material with Young's modulus of 117 GPa, Poisson's ratio of 0.345, and CTE of 17 ppm/°C.

A 56 μm × 56 μm square of the Sn-coated Cu lead frame is modeled. Fifty Sn grains with an average grain size of about 8 μm are considered. The symmetry boundary condition is applied to a lead-frame surface to simulate the fact that the lead frame has a two-sided Sn finish. The Sn finish surface is modeled with the free boundary. The other four sides of the 3D models are modeled with two sides (left and bottom in Fig. 3.12) of symmetry boundary conditions and the other two sides (right and top in Fig. 3.12) of node-coupling boundary condition. A corner node is fixed to eliminate the rigid body motion.

Since the intrinsic compressive stress is normally observed in the Sn film on Cu [1–4], the intrinsic stress is assumed in the FEM model. Following Lau and Pan [2], 8 MPa compressive stress at 20°C is applied to Sn by calling the ANSYS initial stress command ISTRESS. This stress is applied as a hydrostatic pressure, that is, $\sigma_{xx} = \sigma_{yy} = \sigma_{zz} = -8$ MPa and $\sigma_{xy} = \sigma_{yz} = \sigma_{zx} = 0$. This is an approximation and is purely for simplicity reasons. The initial stress of an element is only applied at time zero, and the original value will not be sustained if there is no confinement to that element. In the out-of-plane direction, there is no external constraint after time zero. The hydrostatic pressure kind of initial stress will become an "in-plane biaxial stress" in the Sn film after time zero since the Cu lead frame is the only external confinement to the Sn film. This ANSYS trick helps to simplify the process of resolving the in-plane

TABLE 3.2   Whisker Propensity of Sn on Cu Leads

Plating Bath	Whisker per Lead	Maximum Length (μm)
A	4	15
B	35	65

initial stress in each local elemental coordinate system for each crystal grain, which has different orientation. However, strictly speaking, the intrinsic compressive stress applied in the FEM model of the Sn film may be different than 8 MPa due to the Poisson effect. Thermal loads were studied to mimic the air-to-air temperature cycling test. Six loading steps were applied: 20 to $-55°C$ (Step 1), -55 to $85°C$ (Step 2), 85 to $-55°C$ (Step 3), -55 to $85°C$ (Step 4), 85 to $-55°C$ (Step 5), and -55 to $85°C$ (Step 6).

3.6.5   Simulation Results

Two different plating samples were studied. The Sn finishes of the two lead frames were plated in different baths. The XRD results are tabulated in Table 3.1. The corresponding grain structures were generated to conform to the textural fraction of Table 3.1.

Table 3.2 lists the test results of samples from the two different baths. The test was a temperature cycling accelerated test. The whisker counts and length measurements were done after 1000 cycles of -55 to $85°C$ air-to-air temperature cycling. For Bath-A, the average whisker numbers on each lead of the package is 4 with a maximum whisker length of 15 μm. For Bath-B, the corresponding values are 35 whiskers per lead and a 65 μm maximum whisker length. It will be shown that the calculated volume distribution of SED has a correlation to the whisker propensity. The higher SED value corresponds to higher Sn whisker count and length.

Figures 3.13 and 3.14 are the contour plots of the calculated SED of the two samples after six consecutive temperature load steps. In these figures, the grain patterns overlay the corresponding SED contours. Each grain is indexed by its texture orientation (hkl) followed by an in-plane rotation angle (α in Eq. 3.10 and Fig. 3.1) starting from the x-axis of the lab coordinates. The rotation procedure is described in the previous subsection by Equation 3.10. The plasticity is considered by using the isotropic bilinear hardening rule described earlier. In Figures 3.13 and 3.14, some grains have a high SED and others have a low SED. The maximum SED in Figure 3.14, which corresponds to Bath-B, is 0.371 MPa. The corresponding value for Bath-A in Figure 3.13 is 0.322 MPa. It is also clear that the high SED portion is located in some grains with specific orientations. In Figure 3.13, high SED is in (112) grains. In Figure 3.14, the high SED is in (220) grains and in a (420) grain. This phenomenon is an indication of anisotropy of elasticity and thermal expansion. However, due to the random in-plane rotation (α), not every grain with the same (hkl) is in the worst position to suffer the high SED. An example of Young's modulus spatial distribution around the angle α is studied in [3]. Another important phenomenon that can be observed

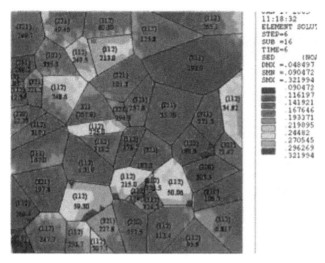

Figure 3.13   Strain energy density of Sn film deposited using Bath-A. The grain orientation of each grain is identified by a texture orientation (hkl) followed by an in-plane rotation angle (α) starting from the x-axis of the lab coordinates. Please visit www.wiley.com/go/Kato/TinWhiskerRisks to access the color version of this figure.

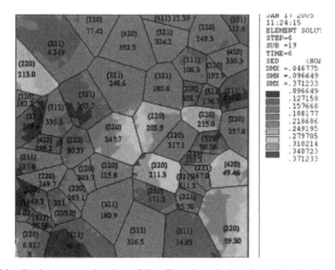

Figure 3.14   Strain energy density of Sn film deposited using Bath-B. The grain orientation of each grain is identified by a texture orientation (hkl) followed by an in-plane rotation angle (α) starting from the x-axis of the lab coordinates. Please visit www.wiley.com/go/Kato/TinWhiskerRisks to access the color version of this figure.

from these figures is that the highest SED is always near the grain boundaries. This explains the fact that Sn whiskers are normally observed at the boundaries of grains experimentally as shown in Figure 3.15.

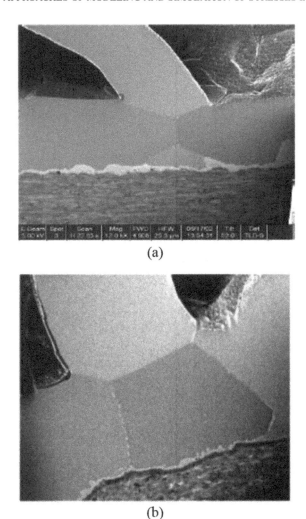

(a)

(b)

Figure 3.15   Examples of Sn whiskers growing on top of grain boundaries. (a) A whisker growing on top of the grain boundary of two Sn grains, (b) a whisker growing on top of two grain boundaries of three Sn grains.

If one assumes that the Sn whisker growth tendency is proportional to the SED and the area under high SED, then a statistical analysis of the calculated SED of both Bath-A and Bath-B may yield more information. Figure 3.16 plots the volume distribution of the SED for both Bath-A and Bath-B. The horizontal axis of Figure 3.16 is the SED. The vertical axis represents the volume fraction with the SED higher or equal to the value of the horizontal coordinate value. For example, in case an SED value on the horizontal axis is 0.3 MPa, on the curve "Plating Bath-A" of the lower panel about 3×10^{-3} (or 0.3%) of the volume of the Sn coating plated in Bath-A has an SED higher than or equal to 0.3 MPa. Corresponding to the curve "Plating Bath-B,"

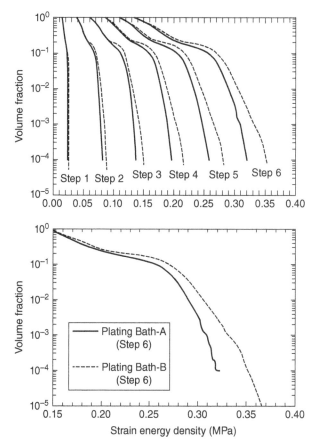

Figure 3.16 Strain energy density distribution comparison for Bath-A and Bath-B. The upper panel is the SED distribution in the first six temperature load steps. The lower panel shows the detail of load Step 6. The horizontal axis is the SED. The vertical axis represents the volume fraction with the SED higher than or equal to the value of the horizontal coordinate value. The solid lines represent Bath-A, and the dashed lines represent Bath-B. More fraction of volume is under higher SED for Bath-B than that of Bath-A for each load step.

about 2×10^{-2} (or 2%) of the Sn has an SED higher than or equal to 0.3 MPa. From Figure 3.16, it can be clearly seen that the Sn finish produced by Bath-B consistently has more volume of Sn material under higher SED than that of Bath-A. From the statistical data of Figure 3.16, there is confidence to believe that the sample with Bath-A will perform better than the sample with Bath-B in the Sn whisker accelerated tests provided the assumption that higher SED promotes whisker growth holds. This is clearly proven by the experimental data listed in Table 3.2.

The upper panel of Figure 3.16 shows the time evolution of the SED of Sn coating from Bath-A and Bath-B. As the temperature cycling progresses, the SED keeps increasing statistically, that is, more and more volume of Sn is under higher SED.

This means that the SED accumulates when temperature cycling progresses. The SED accumulation is a direct manifestation of the Sn plasticity. The SED accumulation explains why Sn whiskers are seen to grow after a certain number of temperature cycles. Although the plasticity model used here is only a first-order approximation, it does explain some of the experimental facts. It is evident that Bath-B consistently has a larger fraction of volume under high SED than that of Bath-A at each loading step of the first six steps. This gap has a tendency of getting larger and larger as temperature cycling progresses. This has a direct correlation with the experimental data in Table 3.2.

If the model only considers the elasticity, SED will be the same as temperature cycling progresses. Figure 3.17 shows the calculated SED results without considering Sn plasticity. The anisotropy of elasticity was considered in the calculation.

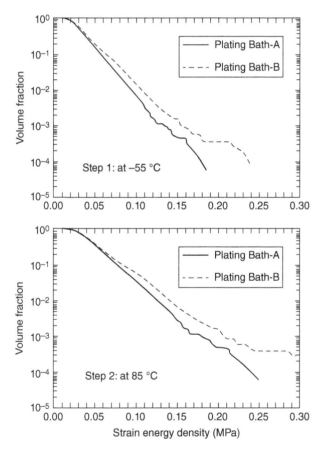

Figure 3.17   Strain energy density distribution when plasticity is not considered. The solid lines are for Bath-A, and the dashed lines are for Bath-B. The thermal loads are from 20 to −55°C (Step 1) and from 20 to 85°C (Step 2). More fraction of volume is under higher SED for Bath-B than that of Bath-A at both temperatures.

The thermal loads are from 20 to $-55°C$ (Step 1) and from 20 to $85°C$ (Step 2). More fraction of volume is under higher SED for Bath-B than that of Bath-A at both temperatures. This is also a good correlation with the experimental data in Table 3.2.

Since the compressive stress is believed to be the driving force of the whisker formation in Sn finish, one may postulate that the hydrostatic stress is responsible for Sn whisker growth. The hydrostatic stress is defined by $(\sigma_1 + \sigma_2 + \sigma_3)/3$, where σ_1, σ_2, and σ_3 are the three principal stresses. Figure 3.18 plots the hydrostatic stress $(\sigma_1 + \sigma_2 + \sigma_3)/3$ distribution when Sn plasticity is not considered. Since yielding is not considered and only pure elastic behavior of Sn is considered, the compressive stress is so large that some areas are more than 150 MPa. Assuming that the whisker growth is the primary stress-relieving mechanism, this kind of plot has a practical

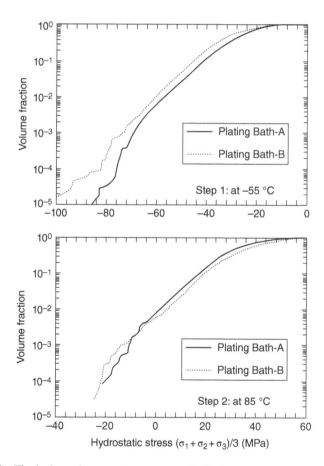

Figure 3.18 The hydrostatic stress $(\sigma_1 + \sigma_2 + \sigma_3)/3$ distribution at two temperatures when the Sn plasticity is not considered. The vertical axis represents the volume fraction with the hydrostatic stress lower than or equal to the value of the horizontal coordinate. The solid lines are for Bath-A and the dashed lines are for Bath-B.

meaning. The higher the compressive hydrostatic stress is (with a negative sign but larger absolute value), the higher the risk of Sn whisker growth will be. The elastic response will not change as the temperature cycling progress. Two distributions (at −55 and at 85°C) represent the two possible stress distributions. It is clearly shown that, at the high temperature end of the thermal cycle (85°C), the two baths show very similar stress behavior. However, at the low temperature end of the thermal cycle (−55°C), more volume fraction of Sn is under higher compressive stress for the Bath-B sample than that of the Bath-A sample. This also agrees well with the test results of Table 3.2, which says that Sn finish produced by Bath-B is more prone to whisker growth.

It should be mentioned that only the elasticity and thermal expansion anisotropy is considered, but not the deemed plasticity anisotropy. The polycrystal plasticity constitutive law is borrowed to account for the single-crystal plasticity, which is the limitation of the current model. As mentioned before, the crystal plasticity theory, which accounts for the effect of slip systems in single crystals, may be the path to breakthrough this limitation. Nevertheless, the SED is insensitive to the exact form of the plasticity constitutive law, which has been proven by the good correlation of the experimental results in Table 3.2 and the simulation results in Figure 3.16. The simplified plasticity model presented here is useful if the SED criterion is adapted for Sn whisker growth.

The creep effect is not considered in the present study. It will be very interesting to see how the creep affects the results. With the crystal plasticity considered, the anisotropy effect of Sn grain is fully accounted. With these mentioned improvements, the current model can be extended to a fully fledged microstructural model for Sn whisker growth prediction. However, even with its current form, if the SED criterion is taken as the Sn whisker growth driving force, the model is validated by experiments and can be used to study whisker growth tendency. The stress-based criterion is more sensitive to the exact form of the plasticity constitutive law. If only anisotropy of elasticity and thermal expansion is considered, using the relative comparison of the stress distributions to predict the whisker growth tendency, it is also attested by the experimental results.

REFERENCES

1. Barsoum MW, Hoffman EN, Doherty RD, Gupta S, Zavalaingos A. Driving force and mechanism for spontaneous metal whisker formation. Phys Rev Lett 2004;93:206104.

2. Lau JH, Pan SH. 3D nonlinear stress analysis of tin whisker initiation on lead-free components. Trans ASME, J Electron Packag 2003;125:621–624.

3. Lee B-Z, Lee DN. Spontaneous growth mechanism of tin whiskers. Acta Mater 1998;46(10):3701–3714.

4. Tu KN. Irreversible process of spontaneous whisker growth in bimetallic Cu–Sn thin-film reactions. Phys Rev B 1994;49(3):2030–2034.

5. JEDEC standard JESD22A121. *Measuring Whisker Growth on Tin and Tin Alloy Surface Finishes*. JEDEC Standards; 2005.

6. JEDEC standard JESD201. *Environmental Acceptance Requirements for Tin Whisker Susceptibility of Tin and Tin Alloy Surface Finishes*. JEDEC Standards; 2006.

7. Ravelo R, Baskes M. Equilibrium and thermodynamic properties of grey, white, and liquid tin. Phys Rev Lett 1997;79(13):2482–2485.

8. Nye JF. *Physical Properties of Crystals*. Oxford: Clarendon Press; 1960.

9. ANSYS 8.1, ANSYS Inc., 2004. [Online]. Available: http://www.ansys.com

10. Lekhnitskii SG. *Theory of Elasticity of an Anisotropic Elastic Body*. San Francisco: Holden-Day; 1963.

11. Anand L. Single-crystal elasto-viscoplasticity: application to texture evolution in polycrystalline metals at large strains. Comput Methods Appl Mech Eng 2004;193:5359–5383.

12. Anand L, Kothari M. A computational procedure for rate independent crystal plasticity. J Mech Phys Solids 1996;44(4):525–558.

13. Lubliner J. *Plasticity Theory*. New York: Collier Macmillan; 1990.

14. Diulin AN, Kirichenko GI, Natsik VD, Soldatov VP. Low-temperature plasticity of Zn-doped β-Sn crystal. J Low Temp Phys 1997;23(10):843–847.

15. Arfken GB, Weber HJ. *Mathematical Methods for Physicists*. 4th ed. Academic Press; 1995.

16. Chen CR, Li SX, Wen JL, Jia WP. Finite element analysis about effects of stiffness distribution on stresses and elastic strain energy near the triple junction in a tricrystal. Mater Sci Eng 2000;A282:170–176.

17. Zhang JM, Xu KW, Ji V. Dependence of stresses on grain orientations in thin polycrystalline films on substrates: an explanation of the relationship between preferred orientations and stresses. Appl Phys Sci 2001;180:1–5.

18. Turki J, Lohe D. Influence of crystalline orientation on the distribution of the load stress in coarse grains. Mater Sci Forum 2002;404–407:477–482.

19. Winther G. Effects of grain orientation dependent microstructure on flow stress anisotropy modeling. Scr Mater 2005;52:995–1000.

20. de Lacy Costtello B, Ratcliffe N. The formation of Voronoi diagrams in chemical and physical systems: experimental findings and theoretical models. Int J Bifurcat Chaos 2004;14(7):2187–2210.

21. Kim D-S, Chung Y-C, Kim JJ, Kim D, Yu K. Voronoi diagram as an analysis tool for spatial properties for ceramics. J Ceram Process Res 2002;3(3):150–152.

22. Vedula VR, Glass SJ, Saylr DM, Rohrer GS, Carter WC, Langer SA, Fuller ER Jr. Residual-stress predictions in polycrystalline alumina. J Am Ceram Soc 2001;84(12):2947–2954.

23. Mathematica 5.1, Wolfram Research, 2004. [Online]. Available: http://www.wolfram.com

4

PROPERTIES AND WHISKER FORMATION BEHAVIOR OF TIN-BASED ALLOY FINISHES

TAKAHIKO KATO

Research & Development Group, Hitachi, Ltd., Tokyo, Japan; Center for Advanced Research of Energy and Materials, Hokkaido University, Sapporo, Japan

ASAO NISHIMURA

Jisso Partners, Inc., Tokyo, Japan

4.1 INTRODUCTION

Adding alloying elements other than lead (Pb) to tin (Sn) finishes can mitigate tin whisker formation, as was reported in the 1960s by Glazunova and Kudryavtsev [1] and Arnold [2]. Although such alloy finishes did not receive much attention after the Sn–Pb finish became standard, the trend toward lead-free soldering that began in the late 1990s has rekindled interest in their use [3–5]. Lead-free tin-based alloy finishes such as tin–bismuth (Sn–Bi), tin–silver (Sn–Ag), and tin–copper (Sn–Cu) are generally recognized as useful. They have since been used in semiconductor devices and other electronic components, mainly by Japanese component manufacturers and their subcontractors. However, with their growing use came negative reports, both experimental and theoretical, on their effectiveness in mitigating whiskers and on their reliability [6–9], resulting in confusion in the market.

This chapter reviews the basic properties of tin-based alloy finishes and the effect of various alloying elements on whisker formation. The focus is on whisker test data

Mitigating Tin Whisker Risks: Theory and Practice, First Edition.
Edited by Takahiko Kato, Carol A. Handwerker, and Jasbir Bath.
© 2016 John Wiley & Sons, Inc. Published 2016 by John Wiley & Sons, Inc.
Companion website: www.wiley.com/go/Kato/TinWhiskerRisks

and potential mechanisms for whisker suppression or enhancement for each element. A close look based on experimental data at the mechanisms of spontaneous whisker formation or suppression in matte Sn–Cu alloy finishes reveals how adding minor elements to the copper base material (lead frame) can significantly change the whisker formation propensity of the alloy finish.

4.2 GENERAL PROPERTIES OF TIN-BASED ALLOY FINISHES (ASAO NISHIMURA)

Three representative lead-free tin-based alloys used on electronic component terminals are Sn–Bi, Sn–Ag, and Sn–Cu with typical compositions, respectively, of 2–3 wt% Bi, 2–3.5 wt% Ag, and 1.5–2 wt% Cu. These alloys were selected not only on the basis of their whisker resistance but also on their electroplatability, solderability, solder joint strength (both with lead-free solder and with conventional Sn–Pb solder), assembly process compatibility, cost-effectiveness, and other properties.

4.2.1 Electroplating of Tin-Based Alloys

The biggest challenge in the electroplating of lead-free tin-based alloys is the large difference in standard electrode potentials between Sn and the codepositing metal. The standard electrode potentials of metals used for tin-based alloy finishes are listed in Table 4.1. For Sn–Pb plating, the Sn and Pb can be codeposited using Sn–Pb anodes in the same manner as for pure metal plating because the standard electrode potentials of Sn and Pb are very close. In contrast, the potentials of alloying elements for lead-free finishes are much more noble than that of Sn. These large differences can cause various problems.

1. The additive metal preferentially deposits into the plating film at low current densities, resulting in variation of alloy composition with current density.
2. The additive metal forms immersion deposits on the tin anode, necessitating frequent replenishment of the additive metal in the form of a liquid concentrate as well as frequent maintenance of the anode.

TABLE 4.1 Standard Electrode Potentials of Metals Used for Tin-Based Alloy Finishes

Electrode	Standard Electrode Potential (V)
Sn/Sn^{2+}	−0.136
Pb/Pb^{2+}	−0.125
Bi/Bi^{3+}	+0.317
Cu/Cu^{2+}	+0.340
Ag/Ag^+	+0.799

3. If the current is cut off while the plated parts are still in the plating bath, the additive metal also forms immersion deposits on the plated layer, leading to reduced solderability or, if the deposits are in the middle of the plated layer, interlayer adhesion failure.

4. The divalent tin (Sn^{2+}) in the electrolyte is oxidized into tetravalent tin (Sn^{4+}) by the more noble additive metal, reducing plating bath stability.

These problems were common in the early stages of mass production of lead-free alloy finishes, resulting in inconsistent plating quality and troublesome bath maintenance [10, 11]. Plating chemical suppliers have since modified electrolyte compositions, thereby reducing the differences between the standard electrode potentials (which has facilitated the codeposition of additive elements) and improving bath stability [11, 12]. Plating companies have been accumulating know-how on bath maintenance through their mass production activities and are now controlling plating conditions more strictly than for Sn–Pb finishes because the alloy composition greatly affects the melting point, whisker formation propensity, and various reliability areas.

4.2.2 Melting Behavior and Solderability

The melting points of Sn–Ag, Sn–Bi, and Sn–Cu finishes are listed in Table 4.2 in comparison to those of pure Sn and Sn–Pb finishes. Since Sn–Ag and Sn–Cu finishes are usually used around their eutectic compositions (Sn–3.5Ag, Sn–0.7Cu), they start melting almost at their eutectic temperature regardless of the precise content of the alloying element. In contrast, Sn–Bi finishes are usually used at compositions far from their eutectic composition (Sn–58Bi), so these finishes have distinct melting onset (solidus) and completion (liquidus) temperatures, both of which depend on the Bi content, as shown in Figure 4.1. A variation in the Bi content of ±1 wt%, for example, causes a variation in the melting onset temperature of about ±5°C. Alloy composition control is thus critical in Sn–Bi plating.

Since the melting onset temperatures of these lead-free finishes are 30–40°C higher than that of the Sn–Pb finish, they may have lower solderability than Sn–Pb

TABLE 4.2 Melting Points of Tin-Based Finishes

Material	Typical Content of Alloying Element for Terminal Finish (wt%)	Melting Point (Solidus Temperature) (°C)	Eutectic Composition	Eutectic Temperature (°C)
Sn–Ag	2–3.5	221	Sn–3.5Ag	221
Sn–Bi	2–3	~210–220[a]	Sn–58Bi	139
Sn–Cu	1.5–2	227	Sn–0.7Cu	227
Sn	—	232	—	—
Sn–Pb	5–40	183	Sn–37Pb	183

[a]Depends on Bi concentration.

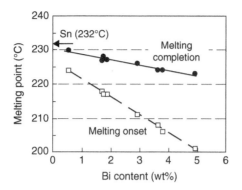

Figure 4.1 Measured melting points of Sn–Bi finish.

when soldered at a temperature lower than their melting point and with a mildly activated flux. While their lower melting points are advantageous, compared to the pure Sn finish, the solder melts first during board assembly, and then the finish film dissolves in the molten solder, meaning that the melting point of the finish is a negligible factor under ordinary soldering conditions. Comparative solderability test data for Sn–Bi-, Sn–Cu-, and Sn–Pb-finished semiconductors with lead-free and with Sn–Pb eutectic solders [13] are shown in Table 4.3. Note that the zero-cross times in the wetting balance test [14] are all less than the generally required 3 s. Semiconductors with either of the lead-free finishes showed sufficient solderability even at the lower soldering temperature.

4.2.3 Solder Joint Reliability

When replacing the conventional Sn–Pb finish with a lead-free one, sufficient reliability must be ensured in combination not only with lead-free solder but also with Sn–Pb solder (backward compatibility). A study conducted by a project group of the Japan Electronics and Information Technology Industries Association (JEITA) [15] measured solder joint pull strengths before and after temperature cycling for low-profile quad-flat-package (LQFP)-type semiconductor packages with various alloy finishes mounted on printed circuit boards using Sn–Ag–Cu and Sn–Pb solders. As shown in Figure 4.2, the pull strength decreased as the number of temperature cycles increased. The Alloy 42 (42% nickel and 58% iron) leads, which had a larger thermal expansion mismatch with the board, showed a greater decrease in strength than the copper (C194) ones. For both types of leads, the decrease was less with the Sn–Ag–Cu solder. The finish material, including the Sn–Pb finish, had little impact on the strength. Component manufacturers have reported [13, 16] that solder joint strengths equivalent to that with the Sn–Pb finish can be obtained with the Sn–Cu finish, for which data are not shown in Figure 4.2.

TABLE 4.3 Solderability Test Data for Sn–Bi, Sn–Cu, and Sn–Pb Finishes

Solder	Sample	Lead Frame	Surface Finish	Wetting Balance Test (Zero-Cross Time (s))			Dip and Look Test (Wetted Area) (%)
				Ave.	Max.	Min.	
Sn–3Ag–0.5Cu (245°C)	QFP 100	Alloy 42	Sn–Cu	0.26	0.28	0.25	≥95
			Sn–Pb	0.19	0.20	0.18	≥95
	LQFP 100	Cu alloy	Sn–Cu	0.64	0.83	0.33	≥95
			Sn–Pb	0.23	0.24	0.22	≥95
	QFP 144	Alloy 42	Sn–Bi	0.27	0.30	0.24	≥95
			Sn–Pb	0.18	0.20	0.16	≥95
		Cu alloy	Sn–Bi	0.42	0.64	0.28	≥95
			Sn–Pb	0.24	0.25	0.23	≥95
Sn–37Pb (230°C)	QFP 100	Alloy 42	Sn–Cu	0.87	1.29	0.39	≥95
			Sn–Pb	0.26	0.29	0.24	≥95
	LQFP 100	Cu alloy	Sn–Cu	1.02	1.46	0.69	≥95
			Sn–Pb	0.28	0.30	0.26	≥95
	QFP 144	Alloy 42	Sn–Bi	0.39	0.60	0.32	≥95
			Sn–Pb	0.22	0.23	0.20	≥95
		Cu alloy	Sn–Bi	1.46	1.65	1.17	≥95
			Sn–Pb	0.26	0.29	0.24	≥95

Preconditioning: steam aging 100°C/100%RH, 4 h.
Flux: WW (water white) rosin flux.
Source: Renesas Electronics Corp. [13].

In the early days of lead-free implementation, when lead-free components were often used in conjunction with Sn–Pb solder, there were concerns about solder joint strength degradation for the combination of the Sn–Bi finish and Sn–Pb solder regardless of evaluation results as those shown in Figure 4.2. This was because the Sn–Pb–Bi ternary alloy can form a eutectic with a melting point of 96°C and a peritectic with a melting point of 135°C. However, it has been demonstrated that Bi content of 5 wt% or less, which is usually used for the surface finish of electronic component terminals, does not form low-melting-point phases or cause strength degradation [17, 18].

4.2.4 Other Properties

Since semiconductor terminal leads are trimmed and formed after plating, surface finish materials that generate less debris, are less likely to stick to the tools, and are resistant to cracking at the lead bend are preferred. The lead-free Sn-based finishes are harder than the Sn–Pb one, so they tend to generate less debris and are less likely to stick to the tools. The Bi and Cu make the deposited film brittle and prone to cracking at the lead bend, as shown in Figure 4.3 [11] for Bi, so their concentrations are usually kept below about 3 wt%. The Sn–Ag finish is advantageous in that a higher concentration of Ag does not result in as much brittleness and does not affect the melting point of the film as much as with Bi and Cu.

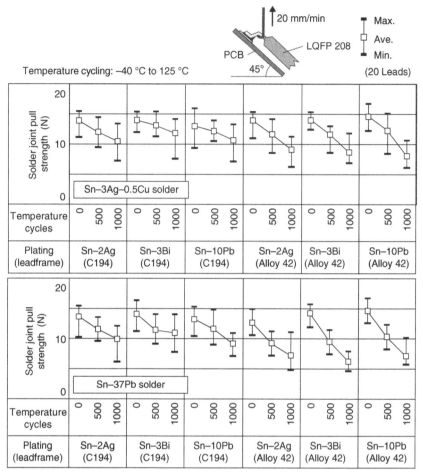

Figure 4.2   Solder joint strengths of various surface finishes before and after temperature cycling. Source: JEITA [15].

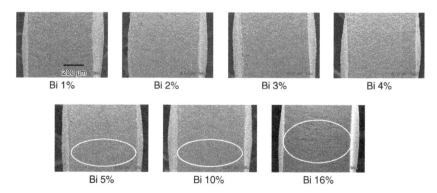

Figure 4.3   Cracking of Sn–Bi finishes after lead forming. Source: Reproduced by permission of the Surface Finishing Society of Japan [11].

TABLE 4.4   Pros and Cons of Lead-Free Tin-Based Alloy Finishes and Pure Tin Finish

Surface Finish	Pros	Cons
Sn–Ag	• Low whisker propensity without heat treatment • Good solderability and mechanical properties	• High cost (special plating equipment, bath/process control, material) • Limited data on whisker propensity and other properties
Sn–Bi	• Low whisker propensity without heat treatment • Good solderability • Proven in practical use since 1998	• Slight toxicity of Bi (very low according to biological tests [19]) • Potential for lower joint strength with Sn–Pb solder (practically no problem with $\leq$5 wt% Bi)
Sn–Cu	• Low environmental impact • No compatibility problem with solder alloys for board assembly	• Weak whisker mitigation effect without heat treatment or other means • Difficulty in measuring Sn–Cu composition on Cu base material
Sn	• Low material/process cost • Ease of bath/process control and maintenance • Low environmental impact • No compatibility problem with solder alloys for board assembly • Widest use for terminal finish	• Need heat treatment or other process for whisker mitigation • Slightly inferior solderability at low temperatures

The pros and cons of the three lead-free finishes and of the pure tin one are summarized in Table 4.4. Although lead-free alloy finishing technologies have been in practical use for electronic component terminals since the early stages of lead-free implementation (late 1990s) and mitigate whisker formation even without heat treatment, they need more complex process control (resulting in higher costs) than the pure Sn finish.

4.3   EFFECT OF ALLOYING ELEMENTS ON WHISKER FORMATION AND MITIGATION (ASAO NISHIMURA)

4.3.1   Whisker Test Data for Tin-Based Alloy Finishes

A study conducted by a JEITA's working group [20, 21] compared the whisker formation and mitigation properties of three lead-free tin-based alloy finishes with those of pure tin and Sn–Pb finishes under various conditions. The configuration of the test specimens is shown in Figure 4.4, and the evaluated specifications are summarized in Table 4.5. The specimens were formed from metal sheets 0.2 mm thick by etching.

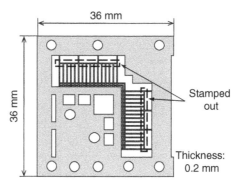

Figure 4.4 Configuration of JEITA lead-frame specimen. Source: Reproduced by permission of the Japan Institute of Electronics Packaging [21].

Tips of the lead-like portions 0.3 mm wide and 5 mm long were stamped out with a die after plating, simulating the lead tips of semiconductor packages. Whisker tests were conducted under three conditions with specimens separately prepared for each test.

- Ambient storage test: uncontrolled office environment, 20,000 h
- High-temperature/humidity storage test: 55°C/85%RH, 10,000 h
- Temperature-cycling test: −40 to +85°C (10-min soak), 2000 cycles

The base material was copper alloy (C194) or Alloy 42. The plating materials were Sn–Ag, Sn–Bi, Sn–Cu, two types of pure Sn [Sn (1) and Sn (2)], and Sn–Pb, all in matte finish. The pure Sn and Sn–Cu finishes on the C194 base material were evaluated with and without heat treatment at 150°C for 1 h within 24 h of plating. The ambient storage test was carried out without preconditioning. The high-temperature/humidity test was carried out for three preconditions: no preconditioning, after Sn–Pb reflow (210°C), and after lead-free reflow (260°C). The temperature-cycling test was carried out for two preconditions: no preconditioning and lead-free reflow. The maximum whisker length was obtained for each sample by periodically inspecting eight predetermined leads with a scanning electron microscope (SEM).

The ambient storage test results are shown in Figure 4.5. The Sn–Cu finish samples without heat treatment could not be evaluated because of damage caused by the lead-tip stamping. Although whiskers 30–50 μm long were observed on the pure Sn finishes without heat treatment, no whiskers were found after 20,000 h ambient storage on the Sn–Ag and Sn–Bi finishes without heat treatment, as well as on the Sn–Cu finish with heat treatment. The "whiskers" found on the heat-treated pure Sn samples after 20,000 h had aspect ratios of less than 2:1, so they are not classified as whiskers according to the definitions given by the International Electrotechnical Commission (IEC) [22] and the JEDEC Solid State Technology Association [23]. Similar evaluation results were reported by the International Electronics Manufacturing Initiative

TABLE 4.5   Specifications for Comparative Evaluation of Whisker Test Specimens

Test	Base Material	Plating Material	Heat Treatment	Reflow Preconditioning
Ambient storage (Office environment)	• C194	• Sn–3.5Ag • Sn–2Bi • Sn (1)[a] • Sn (2)[a] • Sn–10Pb	• None	• None
		• Sn–1.5Cu • Sn (1)[a] • Sn (2)[a]	• 150°C × 1 h	
High-temperature/ humidity storage (55°C/85%RH)	• C194 • Alloy 42	• Sn–3.5Ag • Sn–2Bi • Sn–1.5Cu • Sn (1)[a] • Sn (2)[a] • Sn–10Pb	None	• None • 210°C reflow • 260°C reflow
	• C194	• Sn–1.5Cu • Sn (1)[a] • Sn (2)[a]	• 150°C × 1 h	
Temperature cycling (−40 to 85°C)	• C194	• Sn–3.5Ag • Sn–2Bi • Sn (1)[a] • Sn–10Pb	• None	• None • 260°C reflow
		• Sn–1.5Cu • Sn (1)[a] • Sn (2)[a]	• 150°C × 1 h	
	• Alloy 42	• Sn–3.5Ag • Sn–2Bi • Sn–1.5Cu • Sn (1)[a] • Sn (2)[a] • Sn–10Pb	• None	

[a]Sn (1) and Sn (2) are pure Sn platings with different electrolytes.
Source: JEITA [20, 21].

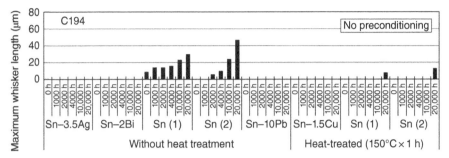

Figure 4.5 Ambient storage test results for various surface finishes. Source: Reproduced by permission of the Japan Institute of Electronics Packaging [21].

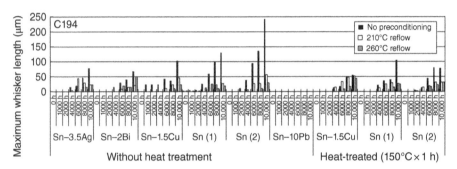

Figure 4.6 High-temperature/humidity storage test results for various surface finishes on copper (C194) base material. Source: Reproduced by permission of the Japan Institute of Electronics Packaging [21].

(iNEMI) [24] for 10,000-h ambient storage: whiskers were not observed on Sn–Ag, Sn–Bi, and Sn–Cu finishes without heat treatment or on pure Sn with heat treatment.

The high-temperature/humidity storage test results are shown in Figures 4.6 and 4.7 for the C194 and Alloy 42 base materials, respectively. The IEC [22] and JEDEC [25] standards require that high-temperature/humidity storage tests be conducted for 2000–4000 h. Within 4000 h, whiskers grew on the lead-free alloy finishes and heat-treated pure Sn finishes on C194 and all the lead-free finishes on Alloy 42 apart from Sn–Bi, but they were shorter than 30 μm, demonstrating the whisker mitigation effect. After 10,000 h, however, all the lead-free finishes on C194 exhibited whiskers longer than 50 μm. These whiskers were found to have grown from the vicinity of corroded areas of the finish film near the lead tip, where the copper base material was exposed, as shown in Figure 4.8. This indicates that neither alloy finishing nor heat treatment sufficiently mitigates whisker growth under conditions that can cause corrosion. Results reported by iNEMI for samples without preconditioning are similar [24].

The results of the temperature-cycling test are shown in Figures 4.9 and 4.10 for the C194 and Alloy 42 base materials, respectively. Only Alloy 42 samples without heat treatment were evaluated since a previous study [26] had shown that heat treatment

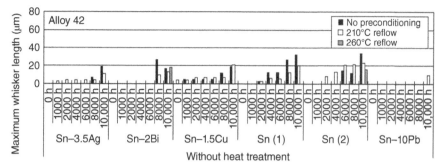

Figure 4.7   High-temperature/humidity storage test results for various surface finishes on Alloy 42 base material. Source: Reproduced by permission of the Japan Institute of Electronics Packaging [21].

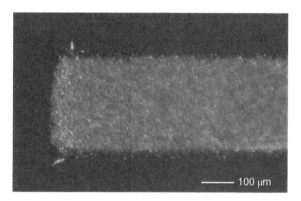

Figure 4.8   Whisker growth from corroded area on lead tip. Source: Reproduced by permission of the Japan Institute of Electronics Packaging [21].

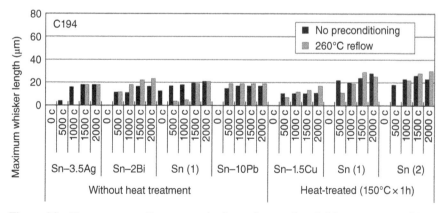

Figure 4.9   Temperature-cycling test results for various surface finishes on copper (C194) base material. Source: Reproduced by permission of the Japan Institute of Electronics Packaging [21].

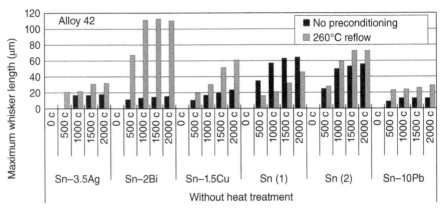

Figure 4.10 Temperature-cycling test results for various surface finishes on Alloy 42 base material. Source: Reproduced by permission of the Japan Institute of Electronics Packaging [21].

has no effect on the temperature-cycling results for Sn and Sn–Cu finishes on Alloy 42. All the C194 samples exhibited whiskers about 20–30 μm long after 2000 cycles, with little effect by the finish material, heat treatment, or reflow preconditioning. On the other hand, the Alloy 42 samples had diverse results. Without preconditioning, whiskers were obviously mitigated with the alloy finishes while long whiskers were observed on the pure Sn finishes. With reflow preconditioning, however, whisker mitigation occurred only for the Sn–Pb and Sn–Ag finishes. Furthermore, whisker growth was enhanced by reflow preconditioning except for the Sn (1) finish, most remarkably for the Sn–Bi finish.

While it is still not fully understood why whisker growth was enhanced by reflow in the temperature cycling of the Alloy 42 base material samples, one probable mechanism has been suggested [20]. Figure 4.11 shows scanning electron micrographs of Alloy 42 specimen surfaces showing the longest whisker (see Fig. 4.10) observed after 2000 temperature cycles [20, 21]. Without preconditioning, the finish films were largely deformed and fragmented by numerous cracks. Cross-sectional observation of the specimens showed that these cracks had reached the interface between the base material and the finish [20]. In contrast, the reflow-preconditioned specimen surfaces were very smooth, showing little sign of deformation and no cracks. Fragmentation of the finish film by cracks should relax the compressive stress and block Sn atom diffusion, resulting in retardation or termination of whisker growth. The finish film subjected to reflow would be more resistant to cracking, facilitating continuous whisker growth. This mechanism also suggests that whisker mitigation by Bi and Cu alloying in temperature cycling can be attributed to embrittlement of the finish film.

As is evident from these reported results, using Sn–Ag and Sn–Bi finishes mitigates whisker formation even without heat treatment. As for the Sn–Cu finish without heat treatment, only a high-temperature/humidity storage test for the C194 specimens, and a high-temperature/humidity storage test and temperature-cycling test for the Alloy 42 specimens were conducted due to specimen preparation problems.

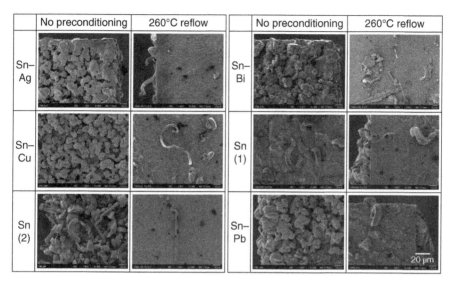

Figure 4.11 Scanning electron micrographs of Alloy 42 specimen surfaces after 2000 temperature cycles. Source: Reproduced by permission of the Japan Institute of Electronics Packaging [21].

While this limited testing showed that the Sn–Cu finish had lower whisker formation propensity than the pure Sn finishes without heat treatment, the whisker mitigation effectiveness of Cu alloying is apparently weaker than that of Ag or Bi alloying, as described in Section 4.3.5.

4.3.2 Whisker Mitigation Mechanism of Tin–Lead (Sn–Pb) Finish

The mechanisms of whisker mitigation when another element is added to tin are not sufficiently understood, even for lead. In fact, they are almost unknown for other elements. Various potential whisker mitigation mechanisms will therefore be reviewed for the Sn–Pb finish before being discussed for lead-free alloy finishes.

The formation of tin whisker is driven by compressive stress generated in the finish film through various mechanisms including Cu–Sn intermetallic compound (IMC) formation, Sn oxidation, and thermal expansion mismatch. The compressive stress is relieved by the Sn atoms diffusing and flowing into surface grains. These grains grow in a manner such that they are pushed out of the film, thereby forming whiskers. Four major mechanisms have been proposed for explaining the effect of lead (Pb) addition on the initiation and growth of whiskers.

1. *Stress relaxation by dispersion of soft phases*   In the Sn–Pb finish film, Pb slightly dissolves in the Sn, forming a solid solution at ordinary temperatures. Most of the Pb precipitates as Pb-rich phases. Since these phases are very soft, they easily deform and relieve the compressive stress in the film. It has also been theorized that compressive stress is relieved by mass transfer of the Pb

because Pb atoms can move more easily within Sn–Pb film than Sn atoms [27, 28]. Although these stress relaxation mechanisms may play a role in whisker mitigation under various conditions including long-term storage and mechanical contact, it seems unlikely that they can explain the whisker mitigation effect when even only 1 wt% Pb [2] or less is added.

2. *Suppression of oxide formation on surfaces of whiskers and/or finish* The addition of Pb to Sn may suppress formation of the oxide film or weaken the oxide film [28–30]. The free surface of the finish without oxide film apparently relieves stress by acting as a sink for point defects and/or dislocations generated by compressive stress in the finish film [30, 31]. If oxide film does not form on the surface of a "whisker" (an extrusion from the finish surface), the "whisker" is considered to easily grow in the lateral direction, leading to growth in hillock, not filament, form [29]. However, data is lacking on how much the presence of Pb suppresses the formation of oxide film. Moreover, whiskers have grown in vacuum even after sputter etching of the oxide film [32]. Thus, the effect of oxide film remains to be fully elucidated.

3. *Effect on grain boundary diffusion of Sn atoms through equiaxialization of grains* Pure Sn finish film usually has a columnar grain structure, with grain boundaries vertical to the film; the film gradually changes into an equiaxed grain structure as Pb is added [8, 33]. Since grain boundaries vertical to the film are subjected to high biaxial in-plane compressive stress, whereas horizontal and oblique grain boundaries are subjected to weaker compressive stress normal to the boundary, Sn atoms flow along the vertical grain boundaries to the horizontal and oblique grain boundaries. In the columnar grain structure, which has few oblique grain boundaries, the Sn atoms concentrate at the limited number of oblique grain boundaries near the surface, thus forming hillocks or whiskers. In the equiaxed grain structure, on the other hand, the entire finish surface is uniformly lifted up due to the many oblique grain boundaries, making formation of hillocks or whiskers difficult [8].

These differences in the grain structure are particularly distinct in bright finishes, which have fine grain sizes in comparison to the finish film thickness, and in finishes with high Pb content. However, in matte finishes, which have grain sizes about equal to the finish film thickness, especially with the commonly used Pb content of 10 wt% or less, differences in the grain structure are not so obvious [33] (see Fig. 4.12 to be explained later). This mechanism is also unlikely to explain the whisker mitigation effect when even only 1 wt% Pb [2] or less is added.

4. *Effect on Cu–Sn IMC formation* When the finish film on the copper base material is cross-sectioned or selectively removed by etching after ambient storage, coarse IMC (Cu_6Sn_5) particles along the Sn grain boundaries are observed at the interface between the pure Sn and base material, whereas uniform IMC growth occurs over the entire interface between the Sn–Pb and base material [20, 21, 28, 33, 34]. Figure 4.12 shows cross sections obtained from JEITA of

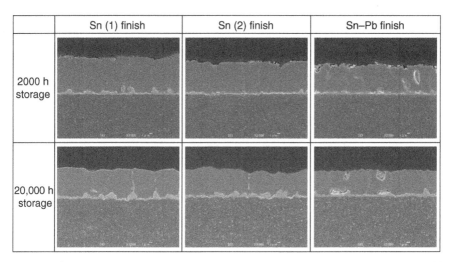

Figure 4.12   Cross sections of pure Sn and Sn–Pb finishes on copper base material after 2000 and 20,000 h ambient storage (Sn (1), Sn (2), and Sn–10Pb specimens without heat treatment or preconditioning, prepared for high-temperature/humidity storage test (see Fig. 4.6)). Source: JEITA [20, 21].

pure Sn [Sn (1) and Sn (2)] and Sn–10Pb finished specimens originally prepared for the high-temperature/humidity storage test (Fig. 4.6) but stored at ambient temperature for 2000 or 20,000 h. IMC grew locally at grain boundaries in the pure Sn specimens, whereas it grew in the form of a thin layer in the Sn–Pb ones.

It is unclear whether the total amount of IMC generated is decreased [30, 34] or unchanged [33] by the addition of Pb. However, since bulk diffusion (lattice diffusion) of Cu into Sn seems to be enhanced by dissolution of Pb in Sn [28, 33], the laminar IMC layer formed through bulk diffusion apparently suppresses the grain boundary diffusion by acting as a barrier against Cu diffusion [28]. Even if the addition of Pb suppresses the stress generated by localized IMC growth, this mechanism cannot explain the whisker mitigation effect of Pb observed under wide-ranging conditions, including temperature cycling and mechanical contact stress [35].

In short, none of these mechanisms explain the whisker mitigation effect of Pb under all conditions. A combination of more than one and possibly other mechanisms might be at work in actual whisker mitigation.

4.3.3   Tin–Bismuth (Sn–Bi) Finish

Sn–Bi was the first lead-free alloy finish put into practical use for electronic component terminals [4] and is widely used by semiconductor manufacturers in Japan and Korea. The whisker mitigation effect of this finish was first reported in 1963 [1], and this effect is now widely recognized due to the growing accumulation of data [3, 20, 21, 26, 36]. Despite its relatively effective whisker mitigation, it has not become widely used among electronic component manufacturers outside Japan and Korea. This is because, as with other alloy finishes, the process is more difficult to control than with pure Sn. Moreover, there were misconceptions about backward compatibility in the early stages of lead-free implementation (see Section 4.2.3) and other reasons.

The whisker mitigation effect of Bi is even less understood than that of Pb. However, the Sn–Bi finish, similar to the Sn–Pb finish, is recognized to suppress nonuniform IMC growth at the interface with the Cu base material [3, 20, 21, 34]. Figure 4.13 shows cross sections of lead-free alloy finished specimens after 2000 and 20,000 h ambient storage, observed in the same way as the pure Sn and Sn–Pb finished specimens (Fig. 4.12) [20, 21]. A uniform IMC layer formed at the interface between the Sn–Bi finish and Cu base material, as with the Sn–Pb finish (Fig. 4.12), and localized IMC growth at the Sn grain boundary was suppressed.

Figures 4.14 and 4.15 show data reported by Baated et al. [34], who stored specimens of various Sn-based finishes on the Cu base material at ambient temperature for 0–7 days, selectively removed the finish films by etching, and observed the morphologies of the exposed IMCs (Fig. 4.14), and measured the volume of the IMCs with a laser microscope (Fig. 4.15). As shown in Figure 4.15, IMC growth in the Sn–Bi finish was suppressed as much as in the Sn–Pb finish, suggesting that Bi has an effect similar to that of Pb. The stress within Sn–Bi finish film was reported to have changed from compressive to tensile with increasing Bi content [37].

These mechanisms can explain the whisker mitigation effect of finishes on Cu base material only under ambient storage conditions, not the effect of Bi addition under

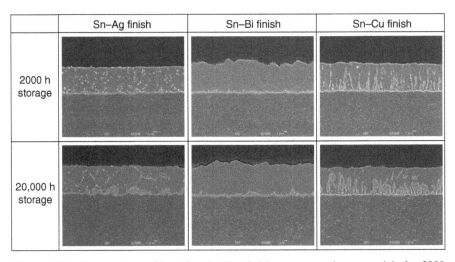

Figure 4.13 Cross sections of lead-free tin alloy finishes on copper base material after 2000 and 20,000 h ambient storage (Sn–3.5Ag, Sn–2Bi, and Sn–1.5Cu specimens without heat treatment or preconditioning, prepared for high-temperature/humidity storage test (see Fig. 4.6)). Source: JEITA [20, 21].

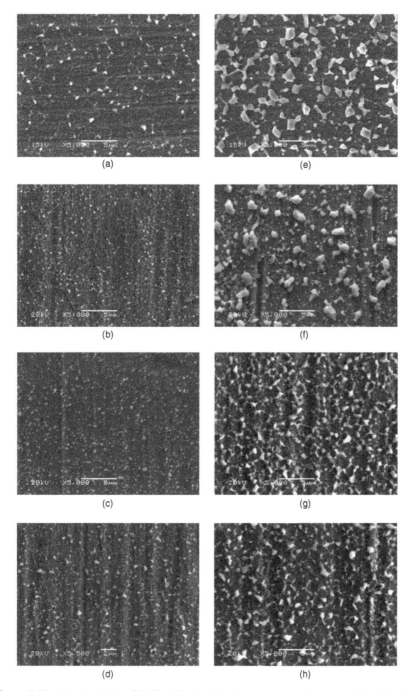

Figure 4.14 Morphologies of IMCs at the interface between copper base material and various tin-based finishes: (a, e) Sn, (b, f) Sn–Ag, (c, g) Sn–Bi, and (d, h) Sn–Pb; (a–d) immediately after plating and (e–h) after 7 days. Source: Reproduced by permission of Springer Science and Business Media [34].

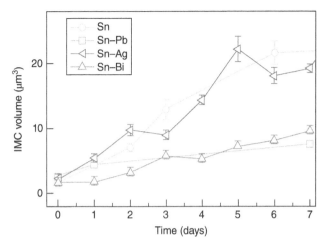

Figure 4.15 Volume change in IMCs formed between copper base material and various tin-based finishes. Source: Reproduced by permission of Springer Science and Business Media [34].

high-temperature/humidity storage and temperature-cycling conditions (see Section 4.3.1) or under mechanical contact stress conditions [35]. The data on the Sn–Ag and Sn–Cu finishes in Figures 4.13–4.15 are explained in the next two sections.

4.3.4 Tin–Silver (Sn–Ag) Finish

The Sn–Ag finish has also been in practical use in Japan since around 2000, its excellent whisker mitigation effect having been demonstrated as shown in Section 4.3.1 and in the literature [3, 20, 21, 24, 38]. However, its use also has not become widespread for surface finishing of component terminal leads outside Japan due to the difficulties in process control, and only limited information has been published on this finish.

When Ag is added to Sn, the deposited film has a fine equiaxed grain structure, which is believed to be beneficial for whisker mitigation [8, 33, 39, 40], as indicated by the cross-sectional photographs in Figure 4.13. The formation of Ag–Sn IMC after electrodeposition has been theorized to cause tensile stress due to volume contraction [37]. Moreover, mutual diffusion of Sn and Cu apparently occurs at the interface between the Cu base material and the Sn–Ag film, which relieves the compressive stress caused by Cu–Sn IMC formation within the deposited layer [38]. However, at the interface between the Cu base material and the Sn–Ag finish, coarse IMC particles similar to those in the pure Sn finish have been found [34], as shown in Figures 4.13 and 4.14.

The reason these coarse particles do not cause compressive stress high enough to form whiskers is not well understood. While ragged coarse IMC particles are formed in the pure Sn finish, those in the Sn–Ag finish are more rounded. This difference in particle shape may affect stress evolution in the finish film. In short, the effect on IMC formation alone cannot explain the effect of Ag addition under high-temperature/humidity storage and temperature-cycling conditions (see Section 4.3.1) or under mechanical contact stress conditions [35].

4.3.5 Tin–Copper (Sn–Cu) Finish

Although there have been a number of reports since the 1960s [1] on the effect of Cu addition, they have been divided over whether it suppresses [1, 2, 26, 41] or enhances [6–8, 10, 42–44] whisker growth.

In the evaluation by Yanada et al. [41], who investigated whisker mitigation under ambient storage conditions with increasing Cu content, IMC formation was better suppressed at the interface between the Sn–Cu finish and Cu base material than with the pure Sn finish, suggesting that codeposited Cu in the finish film suppresses Cu diffusion from the base material. A similar tendency can also be seen in the cross-sectional images in Figure 4.13. They show that a large number of IMC particles precipitated along the grain boundaries of the Sn–Cu finish and agglomerated with time. However, the amount of coarse IMC particles that grew from the base material seems to be less than that in the pure Sn and Sn–Ag finishes.

A number of studies have investigated the whisker enhancement effects of Cu addition, mainly on bright finishes. Boettinger et al. [8], for example, electroplated bright Sn, Sn–Cu, and Sn–Pb films on phosphor–bronze cantilever beams and found that high compressive stress was already present in the Sn–Cu finish soon after plating. They also found that while the pure Sn finish developed only compact conical hillocks within several days after plating, the Sn–Cu finish developed both filamentary whiskers and contorted hillocks.

They attributed the high compressive stress in the finish soon after plating to volume expansion of the finish film caused by the rapid precipitation of Cu_6Sn_5 particles from the supersaturated Sn–Cu solid solution formed by electroplating. They postulated that active surface grains in pure Sn deposits grow into hillocks because they grow laterally due to grain boundary motion as they are pushed upward, whereas these active surface grains in Sn–Cu deposits become filamentary whiskers because of grain boundary pinning by Cu_6Sn_5 precipitation along the grain boundary, which concentrates the growth of the grains in the upward direction.

Pedigo et al. [43] investigated the effect of Cu concentration on deposit stress and surface defect (whisker and/or hillock) morphologies using the same electrolytes and electroplating conditions as Boettinger et al. Building on the work of Pedigo et al. [43], Sarobol et al. [44] investigated the effect of Pb addition and created a "defect phase diagram" showing defect morphology and defect density (number per unit area) as a function of the Cu and Pb concentrations. According to Pedigo et al. [43], the morphologies of hillocks can be depicted schematically, as shown in Figures 4.16 and 4.17, with the morphology dependent on the Cu concentration. The steps in these figures were caused by vertical "uplift" or "extrusion, " the terraces were caused by lateral grain boundary motion, and the ridges indicate trails of triple-line boundaries

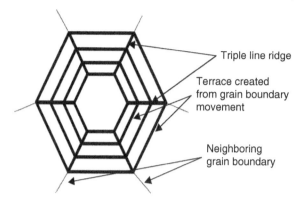

Figure 4.16 Schematic top view of hillock with flat top showing triple-line ridges and terraces due to grain boundary movement. Source: Reproduced by permission of the Institute of Electrical and Electronics Engineers [43].

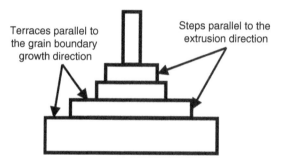

Figure 4.17 Schematic side view of hillock with extruded top showing terraces due to grain boundary movement and steps from extrusion. Source: Reproduced by permission of the Institute of Electrical and Electronics Engineers [43].

between the hillock grain and the neighboring surface grains. The presence of multiple terraces and steps on hillocks suggests the alternating dominance of uplift and grain boundary motion. The morphologies of hillocks are determined by the combination of these two processes.

Pedigo et al. observed hillocks with flat tops, as shown in Figure 4.16, on finishes with low Cu concentrations. When the Cu concentration was increased, they observed hillocks with extruded tops, as shown in Figure 4.17, and whiskers. These observations support the model in which IMC precipitates along grain boundaries impede grain boundary motion. Pedigo et al. also reported increasing compressive stress with an increasing Cu concentration. Sarobol et al. [44] added both Cu and Pb to pure Sn films and observed hillocks with gradual broadening and with few or no defined steps, terraces, and ridges. These observations were summarized by Sarobol et al. [44] into a defect phase diagram (Fig. 4.18).

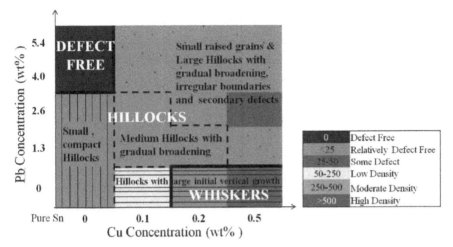

Figure 4.18 Defect phase diagram as a function of Cu and Pb concentrations in bright tin films 240 days after deposition. Source: Reproduced by permission of the Institute of Electrical and Electronics Engineers [44]. Please visit www.wiley.com/go/Kato/TinWhiskerRisks to access the color version of this figure.

The investigations by Boettinger et al. [8], Pedigo et al. [43], and Sarobol et al. [44] focused on bright electroplated finishes, on which hillocks and whiskers are likely to grow soon after plating due to initial high compressive stresses and for which hillock and whisker growth is sensitive to grain boundary movement because of their fine grains. In contrast, with matte finishes, which are commonly used for semiconductor terminals and other components, the addition of Cu to Sn may not enhance whisker propensity in the way they postulated.

Nevertheless, the whisker mitigation effectiveness of Cu seems to be weaker than that of Ag and Bi, even with matte finishes. Hence, Sn–Cu finishes are often used in combination with other mitigation practices such as heat treatment [16] or base material selection [45–52]. The whisker mitigation effectiveness of the heat-treated Sn–Cu finish was shown earlier (Figs. 4.5, 4.6, and 4.9). An example of whisker mitigation by selecting a specific base material for the matte Sn–Cu finish is discussed in the next section.

4.4 DEPENDENCE OF WHISKER PROPENSITY OF MATTE TIN–COPPER FINISH ON COPPER LEAD-FRAME MATERIAL (TAKAHIKO KATO)

This section describes the whisker formation propensity of a matte tin–copper (Sn–Cu) finish and how spontaneous whisker formation can be suppressed by the use of an appropriate copper lead-frame material. It also introduces metallurgical techniques studied by Kato et al. [45–52] in order to illuminate the mechanisms of the formation and suppression of whisker growth from the Sn–Cu coating.

Spontaneous whisker formation from a matte Sn–Cu coating electrodeposited on different Cu lead-frame materials, namely, copper–iron (hereafter, CUFE; corresponding to CDA number C19400) and copper–chromium (CUCR; CDA number C18045), differs substantially. In studies by Kato et al. [45–52] of the effects of long-term storage at room temperature, whiskers did not grow from the Sn–Cu coating on the CUCR lead frame, whereas long whiskers, up to about 200 μm long, grew from that on the CUFE lead frame.

Microstructural field-emission scanning transmission electron microscopy (FE-STEM) and field-emission transmission electron microscopy (FE-TEM) observations and electron-backscatter-diffraction-pattern (EBSP or EBSD) characterizations at vertical cross sections of the Sn–Cu coated lead frames, X-ray diffraction (XRD) stress measurement of the coatings, finite-element analysis (FEA) of coating stress distributions, molecular-dynamics simulation of atom diffusion in the coatings, and investigation of the correlation between the whisker roots and coating microstructures using a planar slicing method were introduced as evaluations for the CUFE and CUCR samples.

4.4.1   Materials and Whisker Propensity

The specifications of the lead-frame samples used in the aforementioned study are listed in Table 4.6, and the ages of the samples are shown in Figure 4.19. One kind of sample had a tin–copper coating on a CUFE lead frame (Sn–Cu/CUFE), and the other kind had a tin–copper coating on a CUCR lead frame (Sn–Cu/CUCR). A 10-μm-thick Sn–Cu coating was electrodeposited under the same conditions on both lead frames, so the coatings had the same copper content (~2 mass%) with the balance being tin.

Matted electrodeposition was used for the Sn–Cu electrodeposition; it was conducted in a commercial fabrication environment with a high current density, and commercial Cu lead-frame materials, CUFE and CUCR, were used. The CUFE (CDA number: C19400) contained three minor elements: 2.4 mass% iron, 0.13 mass% zinc, and 0.08 mass% phosphorous. The CUCR material (CDA number: C18045) contained 0.3 mass% chromium, 0.25 mass% tin, and 0.2 mass% zinc. The lead-frame samples had the same thickness, 0.15 mm, but their widths differed: the Sn–Cu/CUCR samples had a width of 0.3 mm, and the Sn–Cu/CUFE samples, used for the scanning electron microscopy (SEM)/FE-STEM/FE-TEM observations and EBSP measurements, had a width of 0.2 mm.

TABLE 4.6   Samples Used in Study by Kato et al. [45–52]

Lead Sample (Coating/Lead Frame)	Sn–Cu Coating	Commercial Cu Lead-Frame Materialw (Chemical Composition)
(a) Sn–Cu/CUFE	• Electrodeposited[a]	CUFE: CDA number C19400
	• Matted	(Fe: 2.4, Zn: 0.13, P: 0.08, Cu: bal. (mass%))
(b) Sn–Cu/CUCR	• ~2 mass% Cu–bal. Sn	CUCR: CDA number C18045
	• Thickness: 10 μm	(Fe: 2.4, Zn: 0.13, P: 0.08, Cu: bal. (mass%))

[a]Electrodeposition was done in a commercial environment.

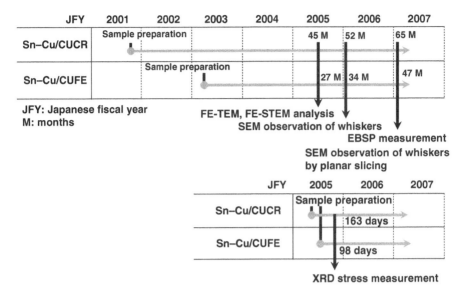

Figure 4.19 Age of samples for each examination (periodic inspection). Two sets of samples were used for each examination. Each examination was conducted during the same inspection period. Sn–Cu/CUCR sample with no whiskers was always older than Sn–Cu/CUFE sample.

Before electrodeposition of the Sn–Cu coating, the lead-frame surfaces were cleaned using a sequential process (chemical polishing, water rinsing, acid dipping) to remove the surface oxide film. The coated samples were stored at room temperature. The examinations listed in Figure 4.19 were conducted during the same period for the two sets of samples. The Sn–Cu/CUCR sample was always older than the Sn–Cu/CUFE sample in each examination.

The whisker initiation tendency was examined by conventional SEM using 120 pieces of lead that had been stored for 34 months as Sn–Cu/CUFE samples and 80 pieces of lead that had been stored for 52 months as Sn–Cu/CUCR samples. Further SEM inspection was done using 20 pieces of lead collected from both sample types.

SEM observation of the same Sn–Cu coating electrodeposited on the two lead-frame materials, CUFE and CUCR, showed significant differences in whisker initiation (Fig. 4.20). No whisker initiation was observed on any of the Sn–Cu/CUCR samples after 52 months of storage while a considerable number of filamentary whiskers (with a maximum length of more than 200 μm) had formed on the Sn–Cu/CUFE samples after a shorter storage time (34 months).

Continuous SEM inspections performed at the same time as the 47-month-old Sn–Cu/CUFE sample examination (Fig. 4.21 (T. Kato, unpublished data)) revealed that whiskers did not grow on any of the Sn–Cu/CUCR samples even for a storage time of 65 months (Fig. 4.22 (T. Kato, unpublished data)). It was qualitatively determined that more whiskers were initiated and grew on the surface of the 47-month-old Sn–Cu/CUFE samples than on that of the 34-month-old Sn–Cu/CUFE samples.

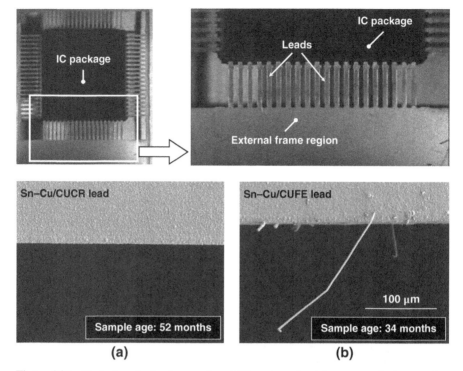

(a)   (b)

Figure 4.20 Typical optical micrographs of IC package lead-frame sample (upper photos) and SEM images of surfaces of two lead-frame samples: (a) Sn–Cu/CUCR and (b) Sn–Cu/CUFE. Source: Reproduced by permission of the Institute of Electrical and Electronics Engineers [52].

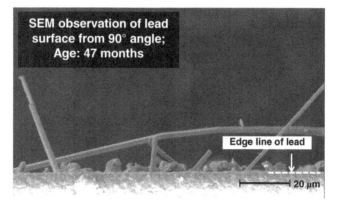

Figure 4.21 SEM image of whiskers on Sn–Cu/CUFE sample (side elevation view of lead).

Figure 4.22 SEM image of breadth-wise surface of Sn–Cu/CUCR lead sample.

4.4.2 Crystalline Orientations and Grain-Size Distributions at Cross Sections of Leads

The coatings were characterized by EBSP measurement [47, 50–52] for vertical cross sections of the longitudinal lead-frame direction for Sn–Cu/CUFE and Sn–Cu/CUCR samples and on the planar-sliced surface of the coating at about the center of the lead-frame width for an Sn–Cu/CUFE sample. This was done using a TexSEM Laboratories orientation imaging microscopy system (MSC-2200) operating at an acceleration voltage of 20 kV. The resolution of the measurement was changed for each EBSP measurement and is indicated by the step size.

Inverse-pole-figure (IPF) maps in the normal direction obtained by EBSP measurements on sample cross sections are shown in Figure 4.23. The Sn–Cu coating clearly had a columnar structure with no oriented grains. On the other hand, the Cu lead-frame materials had cold-rolled textures, indicating that the base Cu lead-frame material did not cause the growth of an epitaxial Sn–Cu coating in either sample type. The grain-size distributions of the coatings and lead-frame materials were also evaluated using EBSP measurements (Fig. 4.24). There was no substantial difference between the grain diameter distributions of the two samples for either the coating or the lead-frame material. It can therefore be concluded that neither the grain orientation of the columnar structures and grain-diameter distributions in both coatings nor the lead-frame materials were the underlying causes of the difference in whisker-initiation tendencies of the two sample types.

The cross-sectional microstructures of the samples were observed with an FE-STEM (Hitachi HD-2000) and an FE-TEM (Hitachi HF-2000) operating at an acceleration voltage of 200 kV using TEM thin foil (about 100 nm thick) trimmed with a focused-ion beam (FIB). These observations made it possible to examine the vertical section of each lead-frame sample (composed of a Cu lead-frame

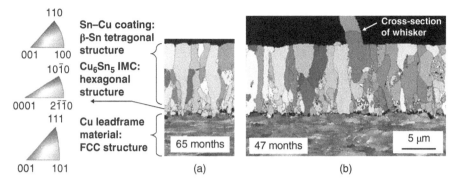

(a)   (b)

Figure 4.23   Crystalline orientations evaluated by EBSP measurement for (a) Sn–Cu/CUCR and (b) Sn–Cu/CUFE samples. Source: Reproduced by permission of the Institute of Electrical and Electronics Engineers [52]. Please visit www.wiley.com/go/Kato/TinWhiskerRisks to access the color version of this figure.

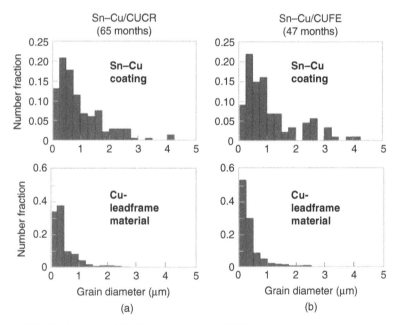

Figure 4.24   Grain-size distributions measured by EBSP. Source: Reproduced by permission of the Institute of Electrical and Electronics Engineers [52].

material and an Sn–Cu coating). The distributions of the minor and major elements of the FIB-processed TEM thin foils were determined using energy dispersive X-ray (EDX) analyzers attached to the FE-STEM (NORAN, Vantage EDX system) and FE-TEM (EDAX, Genesis series). The structures of the IMCs formed at the

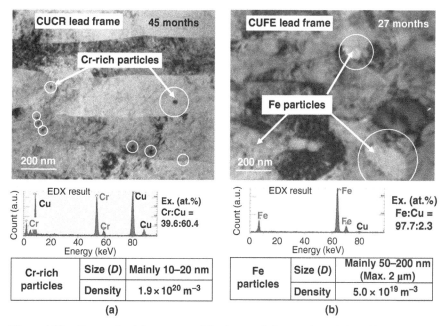

Figure 4.25 Copper lead-frame material characteristics evaluated by FE-TEM/EDX. Source: Reproduced by permission of the Institute of Electrical and Electronics Engineers [52].

interface between the Sn–Cu coating and Cu lead-frame material were evaluated using electron nanodiffraction patterns obtained using a 1-nm-diameter electron beam.

4.4.3 Existence Form of Minor Elements in Lead-Frame Materials

FE-TEM images of the Cu lead-frame materials are shown in Figure 4.25. In the CUCR lead frame, very fine chromium (Cr)-rich particles had been formed with a high density in the matrix and at the grain boundaries. In contrast, in the CUFE lead frame, large, almost pure iron (Fe) particles had been formed with a low density. The Cr-rich particles are thought to have been pure Cr particles because, in the Cu–Cr binary-alloy system, Cu and Cr cannot form a solid solution [53]. The Cu signal in the EDX results for Cr-rich particles is therefore attributed to electron scattering of the inspection beam in the Cu lead-frame material.

4.4.4 Characteristic of Cross-Sectional Microstructures of Leads

Cross-sectional views in the same field of an Sn–Cu/CUFE lead sample on which significant whiskers formed are shown in Figure 4.26. The larger one is an FE-STEM bright field image, and the two smaller are EDX maps. The dotted lines indicate the large-grained intermetallic compound Cu_6Sn_5 (LGIMC) in the Sn–Cu coating

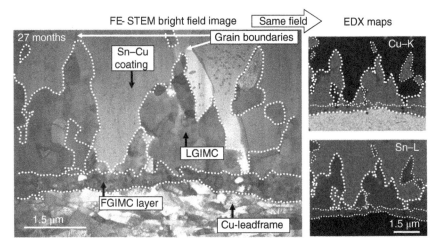

Figure 4.26   Cross-sectional views in same field of lead consisting of Sn–Cu coating on Cu lead frame (CUFE). Dotted trace lines indicate large-grained intermetallic compound (LGIMC) in Sn–Cu coating and fine-grained intermetallic compound (FGIMC) layer. Source: Reproduced by permission of the Institute of Electrical and Electronics Engineers [48].

and the fine-grained intermetallic compound Cu_7Sn_5 (FGIMC) layer. IMC identification is discussed next. The LGIMC built up with a cross-sectional morphology in a wedge-shaped (triangular) structure on the FGIMC layer, which formed with strata at the interface between the Sn–Cu coating and CUFE lead-frame material. The tops of the triangular LGIMCs are located at grain boundaries in the coating. On the other hand, in the Sn–Cu/CUCR sample without any whiskers, the FGIMC formed along grain boundaries with the cross-sectional morphology of a comb-tooth structure in the Sn–Cu coating region, as shown in Figure 4.27.

The IMCs were identified as shown in Figure 4.28. The LGIMCs and FGIMCs both had the same Cu_6Sn_5 hexagonal structure in the electron nanodiffraction patterns, but EDX analysis indicated that the chemical composition of the LGIMC was Cu_6Sn_5, while that of the FGIMC was Cu_7Sn_5. This means that the LGIMC had a hexagonal Cu_6Sn_5 phase and the FGIMC had a hexagonal Cu_7Sn_5 phase in both the Sn–Cu/CUFE and Sn–Cu/CUCR samples. The latter phase is a new one, not previously reported. In 1994, Peplinski et al. [54] identified a $Cu_{6.26}Sn_5$ phase, which has a copper-rich chemical composition, compared with Cu_6Sn_5, by using a single-phase powder sample prepared by diffusing tin from molten tin–lead solder into particles of copper powder.

XRD analysis indicated that the phase had a hexagonal structure. This suggests that, when IMC forms by tin diffusion into copper, it has an excess of copper atoms, compared with the stoichiometric Cu_6Sn_5 compound. The identification of the hexagonal Cu_7Sn_5 phase of FGIMC by Kato et al. [47, 52] is therefore considered reasonable because the phase also forms due to the diffusion of tin atoms from the coating into the copper lead-frame material, as described in Section 4.4.5.

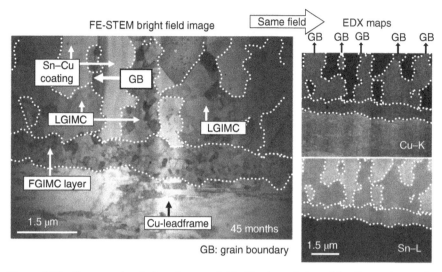

Figure 4.27 Cross-sectional views in same field of lead consisting of Sn–Cu coating on Cu lead frame (CUCR). Dotted trace lines indicate LGIMC in Sn–Cu coating and FGIMC layer. As in the Sn–Cu/CUFE sample, an LGIMC formed on FGIMC layer. Source: Reproduced by permission of the Institute of Electrical and Electronics Engineers [48].

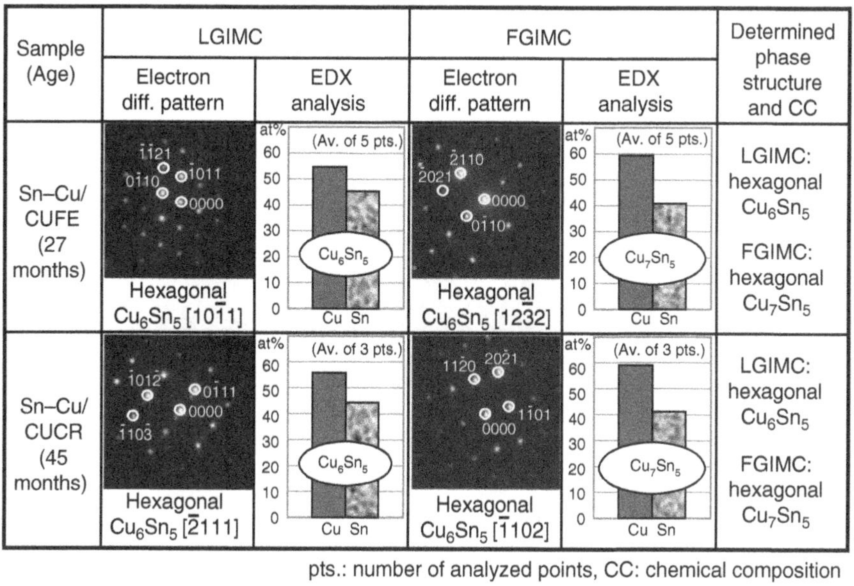

Sample (Age)	LGIMC		FGIMC		Determined phase structure and CC
	Electron diff. pattern	EDX analysis	Electron diff. pattern	EDX analysis	
Sn–Cu/ CUFE (27 months)	Hexagonal Cu$_6$Sn$_5$ [10$\bar{1}$1]	(Av. of 5 pts.) Cu$_6$Sn$_5$	Hexagonal Cu$_6$Sn$_5$ [1232]	(Av. of 5 pts.) Cu$_7$Sn$_5$	LGIMC: hexagonal Cu$_6$Sn$_5$ FGIMC: hexagonal Cu$_7$Sn$_5$
Sn–Cu/ CUCR (45 months)	Hexagonal Cu$_6$Sn$_5$ [$\bar{2}$111]	(Av. of 3 pts.) Cu$_6$Sn$_5$	Hexagonal Cu$_6$Sn$_5$ [$\bar{1}$102]	(Av. of 3 pts.) Cu$_7$Sn$_5$	LGIMC: hexagonal Cu$_6$Sn$_5$ FGIMC: hexagonal Cu$_7$Sn$_5$

pts.: number of analyzed points, CC: chemical composition

Figure 4.28 Identification of IMCs by analysis of electron nanodiffraction patterns and EDX results. Source: Reproduced by permission of the Institute of Electrical and Electronics Engineers [52].

4.4.5   Effect of Minor Elements in Lead-Frame Materials on IMC Formation Morphology

Observation of IMCs by Kato et al. [45–48] showed that the cross-sectional morphology of the LGIMC is a wedge-shaped structure for an Sn–Cu-coated CUFE lead-frame sample and a comb-tooth structure for an Sn–Cu-coated CUCR lead-frame sample, as described in Section 4.4.4. This section describes how minor elements in the lead frames affect the formation morphology of the IMCs.

Figure 4.29 shows the EDX mapping result for Fe atoms in the same cross section as for the results shown in Figure 4.26. The white double and white solid lines trace the LGIMCs and FGIMC layer, respectively. This figure clearly shows that there were Fe particles not only in the Cu lead frame but also in the FGIMC layer. However, there were no particles within the LGIMC and the Sn–Cu coating.

This distribution of Fe particles indicates that the interface between the LGIMC and FGIMC layer was the original surface of the CUFE lead frame before electrode-position of the Sn–Cu coating because Fe particles were initially only in the Cu lead frame (Fig. 4.25). Therefore, the particles can be used as markers to determine the original position of the lead-frame surface. Furthermore, the finding regarding the original surface of Cu lead frame indicates that FGIMC is formed through tin diffusion from the Sn–Cu coating to the Cu lead frame and that LGIMC is formed through copper diffusion from the Cu lead frame to the Sn–Cu coating. That is to say, Sn and Cu atoms mutually diffuse (interdiffuse) across the original surface of the lead frame, and the Fe particles in the Cu lead frame are termed "Kirkendall markers."

The Fe particles in the FGIMC layer were at the bases of the LGIMC triangles, indicating that they suppressed the growth of LGIMC Cu_6Sn_5 by acting as obstacles.

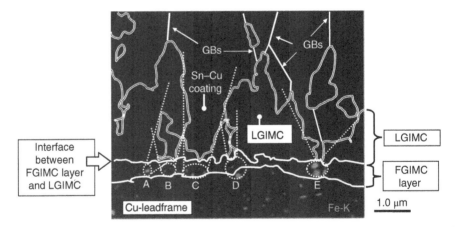

Figure 4.29   EDX mapping of Fe atoms in Sn–Cu/CUFE sample in same cross-sectional views as in Figure 4.26. White double and white solid lines trace LGIMC and FGIMC, respectively. Groupings indicated by white dotted circles (A–E) contain a high density of Fe particles. Sample age was 27 months. Source: Reproduced by permission of the Institute of Electrical and Electronics Engineers [48].

Since they were distributed nonuniformly among several groupings, indicated by the white dotted circles (A–E), containing a high density of Fe particles, they clearly suppressed the growth of LGIMC, as shown by the corresponding white dotted lines. Conversely, LGIMC readily built up in the FGIMC layer immediately above the regions with no Fe particles, as clearly shown in Figure 4.29, resulting in a broad base for the triangular LGIMC. The tops of the LGIMC triangles were always at grain boundaries (GBs) in the Sn–Cu coating. This means that the GBs acted as preferential sites for Cu diffusion from the lead frame to the coating and for LGIMC formation. Therefore, the combination of these two effects (Fe particles and GBs for LGIMC formation) resulted in the wedge-shaped structure of the LGIMC's cross section.

Figure 4.30 shows the result of EDX mapping of Cr atoms in the same cross section as in Figure 4.27. Finely dispersed Cr particles were distributed in the lead frame and in the FGIMC layer. However, no Cr particles were seen in the LGIMC or in the Sn–Cu coating. These results indicate that the interface between the LGIMC and FGIMC layer was the original surface of the CUCR lead frame before electrodeposition of the Sn–Cu coating for a reason similar to that in the case of Fe particles in Sn–Cu/CUFE sample. The Cr particles are thus Kirkendall markers in this case. Segregation of the Cr in the FGIMC layer had no clear presence where the LGIMC growth was suppressed. However, the LGIMC immediately above the FGIMC layer in the Sn–Cu/CUCR sample was relatively narrow, compared with the broad base of the triangular LGIMC in the Sn–Cu/CUFE sample (Fig. 4.26).

This indicates that the Sn–Cu/CUCR FGIMC layer, with its small Cr-rich particles, somewhat suppressed LGIMC growth in the matrix of the Sn–Cu coating in comparison to an FGIMC layer with no Fe particles regions (as indicated in Fig. 4.29). The top of the LGIMC with the comb-tooth structure was always located along a GB

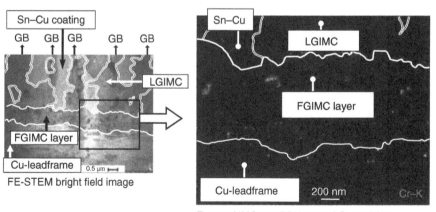

Trace of IMCs on EDX map of Cr particles

Figure 4.30 FE-STEM bright field image and EDX mapping of Cr atoms in Sn–Cu/CUCR sample in same cross-sectional views as in Figure 4.27. LGIMC and FGIMC are traced by white double and white solid lines, respectively. Sample age was 45 months. Source: Reproduced by permission of the Institute of Electrical and Electronics Engineers [48].

of the Sn–Cu coating, similar to that in an Sn–Cu/CUFE sample. (However, this is not evident in Figs. 4.27 and 4.30.) This means that the GBs acted as preferential sites for LGIMC formation in this sample as well. We conclude that, because of their small size, Cr-rich particles have no significant effect on LGIMC growth from the FGIMC layer, unlike Fe particles. Only GBs act as LGIMC formation sites, resulting in the comb-tooth LGIMC structure with the teeth formed vertically on the FGIMC layer in the case of an Sn–Cu/CUCR sample.

The diffusivity of the interdiffusion of Sn and Cu atoms can be evaluated by quantitative comparison of IMC configurations. The thickness of the FGIMC layer formed between the Sn–Cu coating and Cu lead frame in an Sn–Cu/CUFE sample is compared with that in an Sn–Cu/CUCR one in Figure 4.31. That in the latter was more than twice as thick as that in the former although it had been stored longer. This difference in FGIMC layer thickness indicates that the diffusion of Sn atoms from the Sn–Cu coating to the Cu lead frame was suppressed by Fe particles in the Sn–Cu/CUFE sample, whereas it was not strongly suppressed in the Sn–Cu/CUCR sample because of the very fine Cr particles in the CUCR material. The average heights of the LGIMC located along the GBs from the top surface of the FGIMC layers were estimated to be 2870 nm for Sn–Cu/CUFE and 4220 nm for Sn–Cu/CUCR (Fig. 4.32).

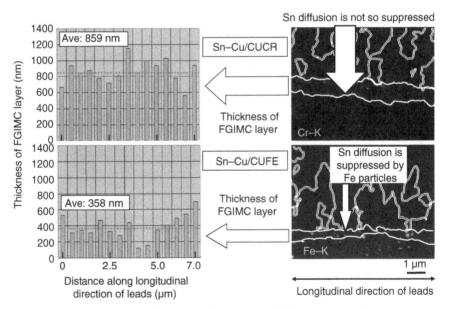

Figure 4.31 Comparison of FGIMC configurations by EDX mapping of Fe atoms in same cross-sectional views as in Figures 4.26 and 4.27. LGIMC and FGIMC are traced by white double and white solid lines, respectively. Sample age was 45 months for Sn–Cu/CUCR and 27 months for Sn–Cu/CUFE. Source: Reproduced by permission of the Institute of Electrical and Electronics Engineers [48].

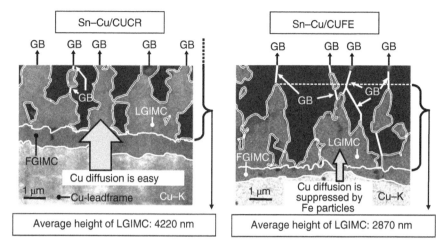

Figure 4.32   Comparison of LGIMC configurations by EDX mapping of Cu atoms in same cross-sectional views as in Figures 4.26 and 4.27. LGIMC and FGIMC are traced by white double and white solid lines, respectively. Sample age was 45 months for Sn–Cu/CUCR and 27 months for Sn–Cu/CUFE. Source: Reproduced by permission of the Institute of Electrical and Electronics Engineers [52].

It is obvious from this result that the apparent Cu diffusivity along GBs in the Sn–Cu coating was also suppressed by the Fe particles in the Sn–Cu/CUFE sample due to the suppression of the supply of Cu atoms from the Cu lead frame to the FGIMC/LGIMC interface. Conversely, Cu atoms easily diffused from the Cu lead frame to the coating through the FGIMC layer in the Sn–Cu/CUCR sample. In the latter case, the fine Cr particles enabled easy diffusion of the Cu atoms from the lead frame to the coating.

This conclusion should be treated with care because the width of the LGIMC immediately above the FGIMC layer in the Sn–Cu/CUCR sample was relatively small, compared with the broad base of the triangular LGIMC in the Sn–Cu/CUFE sample. In explaining this observation, it is reasonable to assume that an Sn–Cu/CUCR FGIMC layer with small Cr particles has less diffusion of Cu atoms to the matrix of the Sn–Cu coating in comparison to an FGIMC layer with no Fe particle regions (indicated in Fig. 4.29). Therefore, the diffusivity of Cu atoms from the FGIMC layer to the matrix of the Sn–Cu coating increases for FGIMC layers in the following order: layer with no Fe particle regions in an Sn–Cu/CUFE sample greater than layer with Cr particles in an Sn–Cu/CUCR sample, which is greater than layer with regions having a high density of Fe particles in an Sn–Cu/CUFE sample.

Finally, a comparison of the FGIMC layer thickness (Fig. 4.31) and LGIMC height along the GBs (Fig. 4.32) showed that the FGIMC layer thickness was less than the LGIMC height along grain boundaries. That is, Cu diffusivity along the grain boundaries seemed to be higher than Sn diffusivity in the Cu lead frame, although no Kirkendall voids, which would have been produced by a difference in diffusivity, were observed at interdiffusion site (interface between LGIMC and FGIMC).

4.4.6 Stress in Coatings and Model for Controlling Whisker Initiation

The data presented earlier clarified that the morphology of the Cu–Sn IMC formed between the coating and Cu lead frame is affected by particles consisting of minor elements in the lead frame. Given the formation morphology of LGIMC Cu_6Sn_5, the difference between the whisker-initiation tendencies of the two sample types (Sn–Cu/CUFE and Sn–Cu/CUCR) was attributed to the difference in the compressive stresses in their coatings. The stresses in the coatings were therefore measured by XRD [45–48], and the results are presented and discussed here. The XRD measurements confirmed that there is a correlation between whisker initiation, LGIMC formation morphology, and compressive stress. On the basis of this correlation, a model for controlling whisker initiation was devised by Kato et al. [45–48].

The residual stresses in the Sn–Cu coatings were measured by XRD using the $\sin^2 \psi$ method (Table 4.7) with $CrK\alpha$ radiation in a RIGAKU MSF-3M XRD system for both sample types. The coatings were electrodeposited under the same conditions as for the samples listed in Table 4.6. The storage times before XRD measurement were 98 and 163 days, respectively, for the Sn/Cu–CUFE and Sn/Cu–CUCR samples. The measurements were performed along two directions: X and Y, the longitudinal and width directions of the lead, respectively. In each case, the surface area where the stress was measured was 5 mm × 4 mm. The peak intensity corresponding to the (312) plane of the β-Sn structure was measured in the range of offset angle $\psi = 0$–$45°$. Thus, the 2θ versus $\sin^2 \psi$ diagram had 11 data points. From the slope in the diagram, the residual stress was estimated using Young's modulus of 43.5 GPa and Poisson's ratio of 0.35 for the electrodeposited Sn–Cu coatings.

A typical 2θ versus $\sin^2 \psi$ diagram presenting results measured by XRD using the $\sin^2 \psi$ method is shown in Figure 4.33. A linear relationship was obtained for both sample types. The stress in the Sn–Cu coatings estimated from the slope of 2θ versus $\sin^2 \psi$ is shown in Figure 4.34. Compressive stresses were induced in both sample types, but that in the Sn–Cu/CUFE one was roughly double that in the Sn–Cu/CUCR one in both the X and Y directions. It is reasonable that many whiskers initiated from the Sn–Cu coating on the CUFE material due to the larger compressive stress in the

TABLE 4.7 Conditions for XRD Measurement of Residual Stresses in Sn–Cu Coating Using $\sin^2 \psi$ Method

Classification in detector scanning plane	Iso-inclination method
Classification in X-ray incident method	Fixed ψ method
Age of samples (days)	Sn–Cu/CUFE: 98, Sn–Cu/CUCR: 163
Stress-measured area	5 × 4 mm
Evaluated peak	(312) plane of β-Sn structure
Offset angle	$\psi = 0$ to $\psi = 45°$
Data in 2θ versus $\sin^2 \psi$ diagram	Eleven data points
Young's modulus, E	43.5 GPa
Poisson's ratio, ν	0.35
Measured direction of lead	X: longitudinal, Y: width

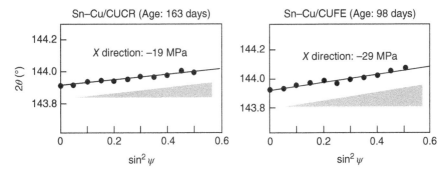

Figure 4.33 Stress data from $\sin^2 \psi - 2\theta$ diagram. Source: Reproduced by permission of the Institute of Electrical and Electronics Engineers [48].

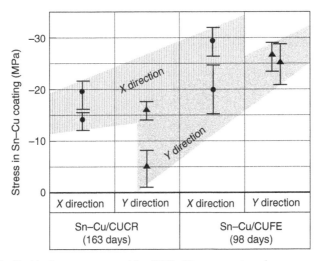

Figure 4.34 Residual stress measured by XRD. Measurements using same samples were done twice for each direction. Error bar for each data point indicates confidence limit. X is longitudinal direction and Y is width direction of lead. Source: Reproduced by permission of the Institute of Electrical and Electronics Engineers [48].

coating. Since no whiskers appeared on the Sn–Cu coating on the CUCR material, which had less compressive stress, there is apparently a threshold for the compressive stress required to initiate whiskers. The threshold lies between the compressive stresses of the Sn–Cu/CUFE and Sn–Cu/CUCR samples.

The difference in the compressive stresses was attributed to the morphology of the Cu_6Sn_5 IMC that formed in the Sn–Cu coatings. Models that explain the interrelations among minor elements, IMC formation morphologies, compressive stresses, and whisker initiation tendencies are presented in Figure 4.35. Minor elements doped into the Cu lead frame produce relatively large Fe particles, ranging from 50 to 200 nm in diameter, in the CUFE lead frame and small Cr particles, 10–20 nm in diameter, in the CUCR lead frame.

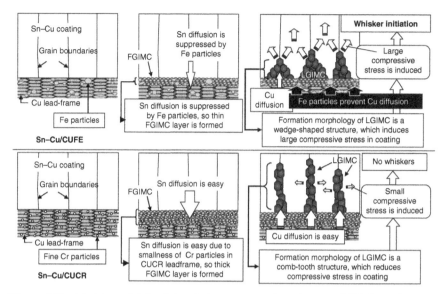

Figure 4.35 Models explaining interrelations among minor elements, IMC formation morphologies, compressive stresses, and whisker initiation tendencies. Source: Reproduced by permission of the Institute of Electrical and Electronics Engineers [48]. Please visit www.wiley.com/go/Kato/TinWhiskerRisks to access the color version of this figure.

During long-term storage, Sn atoms in the Sn–Cu coating diffuse into the lead frame, and, at the same time, Cu atoms in the lead frame diffuse into the Sn–Cu coating preferentially along the grain boundaries in the coating. This interdiffusion of Sn and Cu atoms occurs through the interface between the coating and the lead frame. In the Sn–Cu/CUFE sample, the Fe particles suppressed the diffusion of Sn into the lead frame, resulting in the formation of a thin Cu_7Sn_5 FGIMC layer covering the lead-frame surface. In the Sn–Cu/CUCR sample, the Cr particles had no obvious Sn diffusion suppression effect, so a thick FGIMC layer formed on top of the lead frame, covering it.

Moreover, in the Sn–Cu/CUFE sample, the diffusion of Cu atoms from the lead frame into the coating beyond the FGIMC layer was also suppressed by the Fe particles. Particularly in the regions where their density was high, the Fe particles suppressed the growth of LGIMC Cu_6Sn_5 by acting as obstacles, whereas the Cu atoms diffused somewhat along the GBs, which acted as preferential sites for LGIMC formation. The Fe particles and GBs had a combined effect on LGIMC formation, resulting in an LGIMC configuration with a cross-sectional wedge-shaped structure. Conversely, since the distribution of Fe particles was not homogeneous, the LGIMC growth was kept in the matrix of the Sn–Cu coating immediately above the FGIMC layer with no Fe particles, resulting in the broad base of the triangular LGIMC in the Sn–Cu/CUFE sample.

In the Sn–Cu/CUCR sample, on the other hand, the small Cr particles did not have a significant suppression effect on Cu diffusion, compared with the Fe particles, so marked LGIMC formed along the GBs. It had a comb-tooth cross-sectional structure.

A larger compressive stress was induced in the coating of the Sn–Cu/CUFE sample by the wedge-shaped LGIMC structure, which resulted in considerable whisker initiation but less compressive stress in the coating because the LGIMC's comb-tooth structure could not initiate whiskers on Sn–Cu/CUCR sample. This model can explain all of the results obtained in the study by Kato et al. [45–48]. Controlling the minor elements in Cu lead frames is clearly a key factor in preventing whisker initiation.

4.4.7 Stress Distribution in Coatings

The same studies [47, 49–52] evaluated the stress distribution in the Sn–Cu coating using FEA to precisely determine the contribution of compressive stress to whisker initiation. The configurations of the LGIMCs in the two samples described earlier were taken into account in the FEA calculation. The FEA model is shown in Figure 4.36. The cross section of each sample was assumed to be in a two-dimensional plane-strain condition. Because the model's cross section is symmetrical, it is sufficient to use only the right half in the analysis. In the model, the FGIMC layer and the Cu lead-frame material are omitted, and, to reduce calculation time, the configuration of the LGIMCs is simplified. The boundary condition and mesh structure for the FEA are also shown in the figure.

The calculation of the material constants took into account the bilinear elastic–plastic deformation of the Sn–Cu coating and the elastic deformation of the

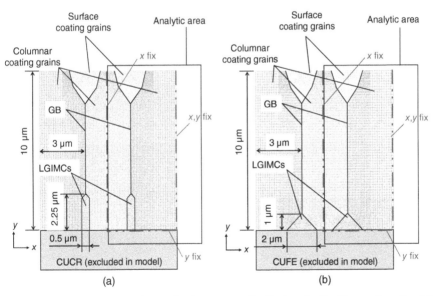

Figure 4.36 FEA models for evaluating stress distribution in Sn–Cu coating for (a) Sn–Cu/CUCR and (b) Sn–Cu/CUFE samples. Cross-sectional shapes of coating grains and LGIMCs are simplified and given on basis of FE-STEM cross-sectional microstructures. Source: Reproduced by permission of the Institute of Electrical and Electronics Engineers [52]. Please visit www.wiley.com/go/Kato/TinWhiskerRisks to access the color version of this figure.

LGIMCs. For the Sn–Cu coating, Young's modulus of 43.5 GPa, Poisson's ratio of 0.35, a yield stress of 30 MPa, and a strain-hardening coefficient of 700 MPa were estimated from bulk solder. For the LGIMCs, Young's modulus of 100 GPa and Poisson's ratio of 0.01 were assumed. Therefore, changes in these values due to crystal orientation were neglected in the analysis, but the same values were set in the models for both samples (Fig. 4.36). The stress distributions in the two samples were calculated assuming the LGIMC growth mechanisms schematically illustrated in Figure 4.37. The initial stress in the coating, including that in the LGIMCs (their initial configuration is shown in gray), was set to zero. The LGIMCs in both cases were assumed to grow by up to about 25% in both the x and y directions, as shown by the dotted lines.

The thermal expansion of the IMC was substituted for the IMC growth in the FEA using the thermal stress analysis function in ABAQUS/Standard. Specifically, the coefficient of thermal expansion (CTE) of the IMC was set to an arbitrary value, and the CTEs of the Sn–Cu coating and Cu lead frame were set to zero. The FEA model was then subjected to a thermal load produced by a uniform rise in temperature.

The FEA results for the distribution of x-directional stress, σ_x, in the Sn–Cu coating, where σ_x is always normal to the GBs, are shown in Figure 4.38. The scale of the two diagrams is the same as that of those in Figure 4.36. Both samples clearly had a two-directional stress gradient: one toward the surface and one toward the base lead-frame material.

The one toward the surface in the Sn–Cu/CUCR sample was smaller than that in the Sn–Cu/CUFE sample. The σ_x along the grain boundary was calculated and is

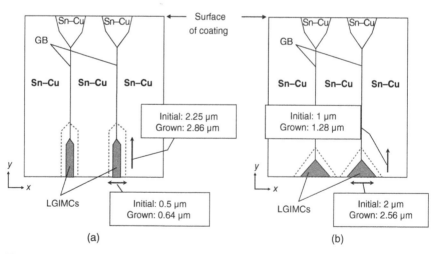

Figure 4.37 LGIMC growth models used in FEA analyses for (a) Sn–Cu/CUCR and (b) Sn–Cu/CUFE samples. Initial stress in coating, including that in LGIMCs (the initial configuration of which is shown in gray), was set to zero. The LGIMCs in both cases were assumed to grow by up to about 25% in both the x and y directions, as shown by dotted lines. Source: Reproduced by permission of the Institute of Electrical and Electronics Engineers [52].

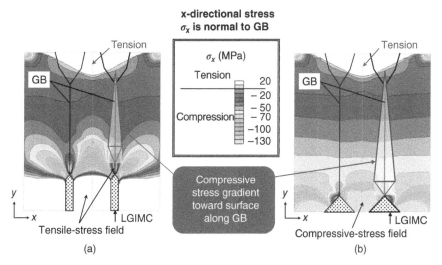

Figure 4.38 FEA results of x-directional stress distribution in coating for (a) Sn–Cu/CUCR and (b) Sn–Cu/CUFE samples. Gradient of compressive stress normal to GBs is indicated by triangles along GBs. Source: Reproduced by permission of the Institute of Electrical and Electronics Engineers [52]. Please visit www.wiley.com/go/Kato/TinWhiskerRisks to access the color version of this figure.

plotted against distance A to C or A to C′ in Figure 4.39. The scale of the two diagrams on the left is the same as that of those in Figures 4.36 and 4.38. The graph shows that the stress gradient along the grain boundary (i.e., the slope) toward the surface in the Sn–Cu/CUCR sample was smaller than that in the Sn–Cu/CUFE sample.

The relationship between the stress gradient and the atom flux along the GB is given by Equation 4.1 [55] and Equation 4.2 [56]

$$J_B = -\frac{D_B}{kT}\frac{\delta\mu}{\delta s} \tag{4.1}$$

$$\mu = \mu_0 - \sigma_n\Omega \tag{4.2}$$

where J_B is atom flux, D_B is the coefficient of GB diffusion, k is Boltzmann's constant, T is absolute temperature, μ_0 is the reference value of the chemical potential, σ_n is stress normal to the GB, Ω is atomic volume, and s is the distance along the GB. According to Equations 4.1 and 4.2, the atom flux along the GB is proportional to the gradient of stress normal to the GB, as schematically illustrated in Figure 4.40. The atom flux toward the surface in the Sn–Cu/CUFE sample is therefore thought to have been larger than that in the Sn–Cu/CUCR sample because the stress gradient toward the surface in the Sn–Cu/CUFE sample was larger than that in the Sn–Cu/CUCR sample (Figs. 4.38 and 4.39).

These findings indicate that the difference between the whisker initiation tendencies of the two samples (Figs. 4.20–4.22) is caused by the difference in

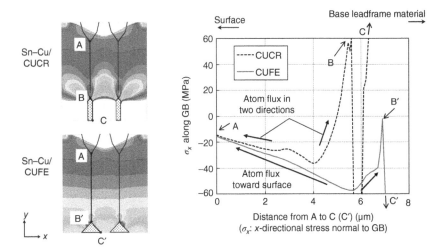

Figure 4.39 Coating depth dependence of stress normal to GBs (calculated by FEA). Source: Reproduced by permission of the Institute of Electrical and Electronics Engineers [52]. Please visit www.wiley.com/go/Kato/TinWhiskerRisks to access the color version of this figure.

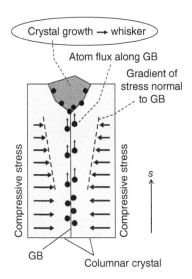

Figure 4.40 Conceptual relationship between gradient of compressive stress normal to GB and atom flux along GB. Whisker initiation is thought be due to crystal growth of surface grain(s) induced by diffusion of atoms along GB. Source: Reproduced by permission of the Institute of Electrical and Electronics Engineers [52].

the amounts of atom flux toward the surface along the GB, which is due to the compressive-stress gradient mentioned in this section. They also suggest that a whisker in the Sn–Cu/CUFE sample could have initiated from a surface grain immediately above a GB or from surface grains located on both sides of the same GB.

The FEA results presented in Figure 4.38 show that, besides a compressive-stress region, a tensile-stress region existed near the surface. The effect of this tensile-stress region on whisker formation should be considered because studies [57–62] have shown that whiskers are formed only by compressive stress induced by the formation of IMC Cu_6Sn_5 between the Sn-system coating and Cu substrate as a result of preferential Cu diffusion from the substrate to the coating along the GBs in the coating. The existence of a stress gradient (Fig. 4.38) means that the atomic lattice expands continuously from the compressive-stress region to the tensile-stress region, as schematically shown in Figure 4.41. Consequently, the elastic lattice strain field could cause Sn atoms to diffuse smoothly through the interface between these two stress regions (since no elastic barrier exists at the interface). Kato et al. [52] therefore concluded that the tensile-stress region has no effect on tin-atom diffusion and, as a result, whiskers can initiate there.

This conclusion agrees with the XRD data reported by Sobiech et al. [63], who showed that, as long as the planar stress gradient was positive (increasingly tensile) toward the surface, Sn whiskers grow at ambient temperature from an as-deposited film with planar tensile stress, although their stress-depth profiles were limited to the near-surface region (depth up to 1 μm).

4.4.8 Tin Diffusion Calculated by Molecular Dynamics

The contribution of compressive stress to whisker initiation was clarified by calculating the correlation between Sn-diffusion sites and whisker-formation sites using molecular-dynamics simulation of atom diffusion and by investigating the correlation between whisker roots and microstructures in the Sn–Cu coating using a planar slicing method [47, 49–52]. The grain boundary (GB) diffusion was compared with

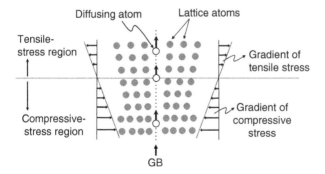

Figure 4.41 Schematic illustration of atom diffusion along GB from compressive-stress region to tensile-stress region in crystal with elastic distortion. Source: Reproduced by permission of the Institute of Electrical and Electronics Engineers [52].

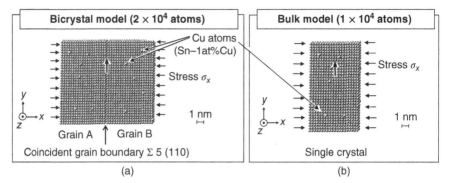

Figure 4.42 Models of atom diffusion for molecular-dynamics simulations. Source: Reproduced by permission of the Institute of Electrical and Electronics Engineers [52].

bulk diffusion by performing a simple, in-house software simulation of atom diffusion using molecular dynamics.

The two models used in the simulation are shown in Figure 4.42. To simulate Sn diffusion at the GB, a bicrystal model with a Σ5 (110) coincident GB was used (a). This model contains 20,000 atoms, of which 200 Cu atoms, that is, 1 at.%, were randomly placed at lattice points. A single-crystal model (b) was used to simulate bulk diffusion. The stress dependence of the GB-diffusion coefficient was investigated by applying stress in the horizontal direction (as indicated in Fig. 4.42).

In the calculations, atomic diffusion was estimated by solving Newton's equation of motion:

$$\frac{m_i \mathrm{d}^2 r_i}{\mathrm{d}t^2} = -\frac{\partial(\Sigma_{i<j} u_{ij} + \Sigma_i \phi_i)}{\partial r_i} \tag{4.3}$$

where m_i and r_i are the atomic mass and atomic position of the ith atom, respectively, t is time, u_{ij} is the interatomic interaction between the ith and jth atoms, and ϕ_i is the potential energy of the ith atom itself [64]. Periodic boundary conditions were applied in the x, y, and z directions. The diffusion coefficient for the y direction was then calculated from the Einstein–Smoluchowski relation:

$$D = \lim_{t \to \infty} D(t) \tag{4.4}$$

$$D(t) = \frac{\langle [y_i(t + t_0) - y_i(t_0)]^2 \rangle}{2t} \tag{4.5}$$

The $y_i(t+t_0) - y_i(t_0)$ represents the displacement of the ith atom in the y direction from time t_0 to $t + t_0$. The angle brackets ($\langle \ \rangle$) in Equation 4.5 indicate the average over all the atoms related to the diffusion. In Equation 4.4, time is set to infinity, but $D(t)$ converged to D in finite time in the simulation. Therefore, we can obtain diffusion coefficient D in finite time. The simulation results for GB diffusion and bulk diffusion are shown in Figure 4.43. The GB diffusion coefficients were clearly much larger than the bulk-diffusion ones across the temperature range. It was therefore concluded that GB diffusion is a more important factor in whisker growth than bulk diffusion.

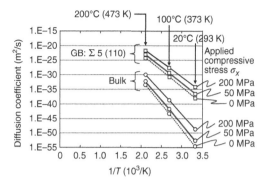

Figure 4.43 Diffusion coefficient at GB and in bulk lattice from molecular dynamics simulation. Source: Reproduced by permission of the Institute of Electrical and Electronics Engineers [52].

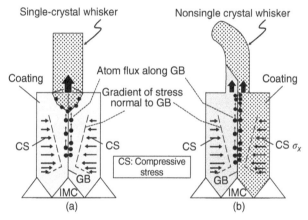

Figure 4.44 Schematic illustrations of whisker-initiation model derived from simulation results in Figure 4.43. A whisker preferentially initiates from surface grain immediately above GB (a) or from surface grains on both sides of same GB in coating (b). Source: Reproduced by permission of the Institute of Electrical and Electronics Engineers [52].

It can be assumed from the simulation results that whiskers preferentially initiate from a surface grain immediately above a GB or from surface grains on both sides of the same GB in the coating, as schematically illustrated in Figure 4.44. This assumption about whisker-initiation sites was verified, as described in the next section, by using an Sn–Cu/CUFE sample to investigate the correlation between whisker roots and coating microstructures.

4.4.9 Planar Slicing Method

The correlation between whisker roots and coating microstructures in the Sn–Cu/CUFE sample was investigated [47, 49–52] by using a planar slicing method [65, 66]. The locations of the roots were determined using SEM panoramic

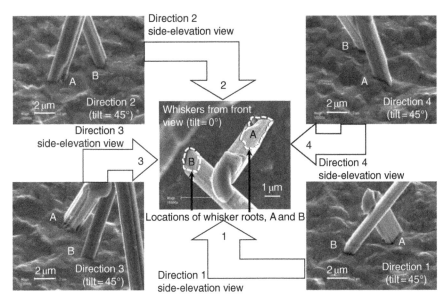

Figure 4.45 SEM images of whiskers from various angles for Sn–Cu/CUFE sample stored for 47 months. Source: Reproduced by permission of the Institute of Electrical and Electronics Engineers [52].

observations of the whiskers. The central image in Figure 4.45 shows a front view of the roots of two whiskers, A and B, in the Sn–Cu/CUFE sample, and the surrounding images show panoramic views of the same whiskers. The whisker root locations were determined from the photographs and are marked in the central image.

To confirm the correlation between whisker roots and coating microstructure, the sample shown in Figure 4.46 was sliced horizontally using a planar slicing method. First, the whiskers were fixed and protected by pouring a resin coating on the surface, as shown in Figure 4.47. The whiskers, along with the resin, were then sliced. The resin was transparent, so the whisker roots were visible through the thinned resin coating (as shown in the figure on the far right). Next, the sample was thinned to a depth of 1 μm from the Sn–Cu-coated surface (Fig. 4.48a).

This made it possible to determine the correlation between the whisker root locations and the coating microstructure located immediately below the roots, as shown in Figure 4.48b. EBSP measurement (Fig. 4.48c) of the same horizontal cross section of the coating as in the top image shows more clearly that the roots were located at GB intersections. It can therefore be concluded that the whisker-initiation site was either a grain located immediately above a GB or surface grains located on both sides of the same GB in the coating.

Finally, the Sn–Cu coating was completely removed by dissolving it chemically (Fig. 4.49a). Figure 4.49b shows the root locations of the same whiskers, A and B, for the same horizontal observation area as in Figure 4.48b and c. The same observation area was achieved for Figures 4.45–4.49 by using FIB markers, which were placed

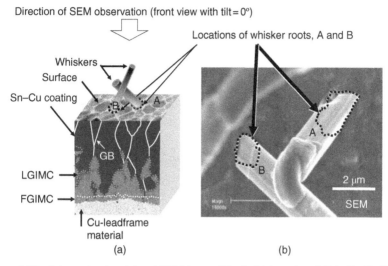

Figure 4.46   Schematic view (a) and SEM image (b) of whiskers A and B in Sn–Cu/CUFE sample stored for 47 months before planar slicing. Source: Reproduced by permission of the Institute of Electrical and Electronics Engineers [52].

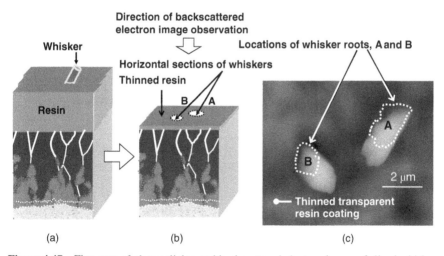

Figure 4.47   First step of planar slicing and backscattered electron image of sliced whisker roots, A and B, in Sn–Cu/CUFE sample stored for 47 months: (a) schematic view of whiskers protected by resin on surface, (b) schematic view of whiskers sliced by mechanical reduction of resin, and (c) backscattered electron image of front view corresponding to (b). Source: Reproduced by permission of the Institute of Electrical and Electronics Engineers [52].

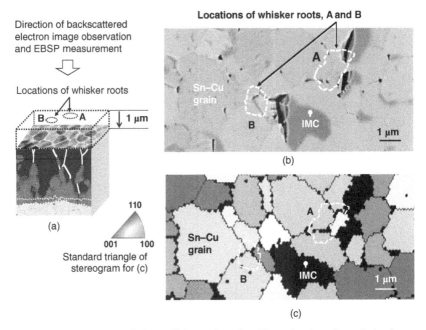

Figure 4.48 Second step of planar slicing and results of investigation of correlation between whisker roots and coating microstructure: (a) schematic view after thinning of Sn–Cu coated surface, (b) backscattered electron image of coating microstructure, and (c) EBSP inverse-pole-figure map with normal direction corresponding to microstructure (b). Step size of EBSP measurement was 0.15 μm. Source: Reproduced by permission of the Institute of Electrical and Electronics Engineers [52]. Please visit www.wiley.com/go/Kato/TinWhiskerRisks to access the color version of this figure.

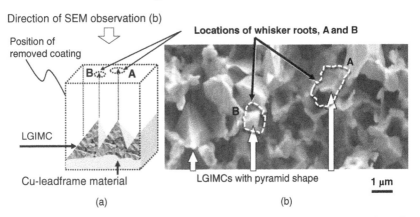

Figure 4.49 Third step of planar slicing and SEM observation of LGIMCs from front view: (a) schematic view and (b) SEM image of LGIMCs from front view after complete removal of Sn–Cu coating by chemical dissolution. Whisker roots A and B are the same as in Figures 4.45–4.48. Source: Reproduced by permission of the Institute of Electrical and Electronics Engineers [52].

around the target area (although the markers are not visible in the figures). This made it possible to get a front view of the shapes of the LGIMCs. Figure 4.49b shows that the whisker roots were located above the peaks and ridgelines of the LGIMC pyramids.

The results presented in the previous sections can be summarized as follows. Whisker-initiation sites are correlated with tin-diffusion sites (i.e., GBs). This correlation is explained by the finding that whisker roots were located above the GB intersections (as determined by planar slicing) and that the dominant tin-diffusion sites were GBs when compressive stress was applied in the normal direction (as determined by molecular-dynamics calculation).

Whisker-initiation sites were located above the peaks of the LGIMC pyramids (Fig. 4.49). Therefore, given the FEA results (Figs. 4.38 and 4.39) and X-ray stress measurement results (Figs. 4.33 and 4.34) presented earlier, it is reasonable to assume that the LGIMC pyramids induced a large compressive stress field above themselves, resulting in enhanced tin diffusion at the GBs immediately above the pyramids. As a result of the compressive-stress field normal to the GBs induced by the LGIMC pyramid, each whisker initiated from either a grain located immediately above a GB or from surface grains located on both sides of the same GB.

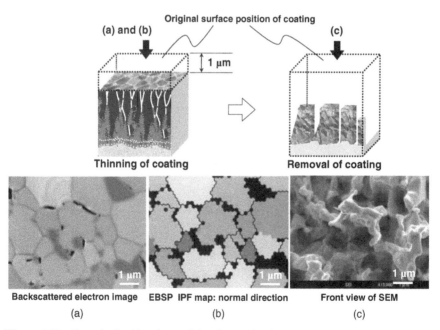

Figure 4.50 Data obtained by planar slicing for Sn–Cu/CUCR sample stored for 65 months. (a), (b), and (c) have the same observation area, which is indicated by the black arrows in the upper schematic views. Step size of EBSP measurement for (b) was 0.2 μm. Source: Reproduced by permission of the Institute of Electrical and Electronics Engineers [52]. Please visit www.wiley.com/go/Kato/TinWhiskerRisks to access the color version of this figure.

Figure 4.50 shows a backscattered electron image (a) and the corresponding EBSP IPF map (b) of the coating microstructure of the Sn–Cu/CUCR sample for reference after the sample was horizontally thinned to a depth of 1 μm from the Sn–Cu coated surface by planar slicing. The figure also shows a frontal SEM view of the LGIMCs (c) when the Sn–Cu coating was completely removed by chemically dissolving it. The LGIMCs clearly did not have a pyramidal shape; instead, they had an almost rectangular shape. With this LGIMC structure, the compressive-stress gradient along the GBs in the coating was small (as shown by the FEA calculations).

4.4.10 Whiskers with Non-Single-Crystal Structure

The Sn–Cu/CUFE sample exhibited whiskers having a bicrystal structure, as shown in Figure 4.23b. Other non-single-crystal whiskers were also easily observed in the same sample, as shown in Figure 4.51, which shows SEM images of three whiskers (A, B, and C) formed after sample storage for 47 months. EBSP IPF maps of the horizontal cross sections of the whisker roots are also shown. It is clear from the maps that the root of whisker B had a single-crystal structure, while those of A and C had bicrystal structures.

Figure 4.52 shows the correlation between the locations of whisker roots B and C (the same ones as indicated in Fig. 4.51) and the coating microstructure immediately below the roots. EBSP measurement of the horizontal cross sections of the coatings

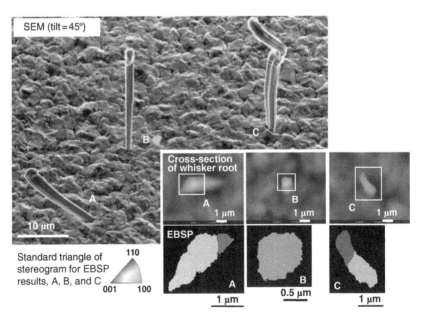

Figure 4.51 Whiskers with non-single-crystal structure in Sn–Cu/CUFE sample stored for 47 months. Step size of EBSP measurement for whisker roots was 0.03 μm. Source: Reproduced by permission of the Institute of Electrical and Electronics Engineers [52]. Please visit www.wiley.com/go/Kato/TinWhiskerRisks to access the color version of this figure.

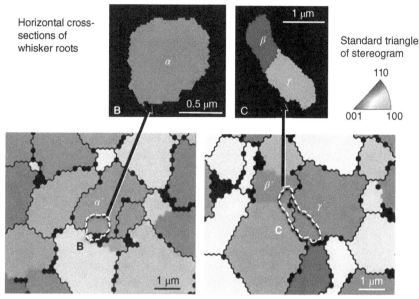

Horizontal cross-sections of whisker roots

Standard triangle of stereogram

Horizontal cross-sections of Sn–Cu coating under whiskers (EBSP IPF map: normal direction)

Figure 4.52 Crystalline orientations of whiskers and Sn–Cu coating under whiskers in Sn–Cu/CUFE sample stored for 47 months. Whiskers B and C are the same as in Figure 4.51. Source: Reproduced by permission of the Institute of Electrical and Electronics Engineers [52]. Please visit www.wiley.com/go/Kato/TinWhiskerRisks to access the color version of this figure.

showed that the roots of whiskers B and C were clearly located on the GBs of the coating. Moreover, the grain orientation α of the root of whisker B was very similar to grain orientation α' of the upper grain of the coating, and the grain orientations β and γ of the root of whisker C were similar to grain orientations β' and γ' of the coating, respectively. These findings support the conclusion that a whisker initiates either from a surface grain located immediately above a GB (Fig. 4.44a) or from surface grains located on both sides of the same GB (Fig. 4.44b).

4.4.11 Summary of Dependence of Whisker Propensity of Matte Tin–Copper Finish on Copper Lead-Frame Material

This section presented four main findings. First, the considerable difference between the whisker-initiation tendencies of matte Sn–Cu coating on CUCR (CDA number C18045) lead-frame material (Sn–Cu/CUCR sample) and Sn–Cu coating on CUFE (CDA number C19400) lead-frame material (Sn–Cu/CUFE sample), which is explained by the correlation between whisker formation, microstructure, and stress. Second, this correlation is supported by FEA of the stress gradient in the Sn–Cu coating. Third, in the Sn–Cu/CUFE sample, the whisker-initiation sites were located above the GBs, including GB intersections, in the Sn–Cu coating, and above the peaks of the LGIMC pyramids. Fourth, these locations coincide with findings by molecular dynamics simulation that the dominant sites for Sn diffusion are GBs.

4.5 CONCLUSIONS

The basic properties of tin-based alloy finishes and the effect of various alloying elements on whisker formation were overviewed. The focus is on whisker test data and potential mechanisms for whisker suppression or enhancement for each element.

A close look based on experimental data at the mechanisms of spontaneous whisker formation or suppression in matte Sn–Cu alloy finishes revealed how adding minor elements to the copper base material (lead frame) can significantly change the whisker formation propensity of the alloy finish.

Significantly different tendencies of whisker formation from the same matte Sn–Cu coating electrodeposited on two different copper lead frames, namely, copper–iron (hereafter, CUFE; corresponding to CDA number C19400) and copper–chromium (CUCR; CDA number C18045), were found.

After long-term storage at room temperature, no whisker formation occurred from the Sn–Cu coating on the CUCR lead frame, whereas long whiskers with a maximum length of more than 200 μm were formed from the Sn–Cu coating on the CUFE lead frame.

Microstructural FE-STEM/FE-TEM/EBSP characterizations at vertical cross sections of the Sn–Cu-coated lead frames, an XRD stress measurement of the coatings, an FEA analysis of coating stress distributions, a molecular-dynamics simulation of atom diffusion in the coating, and an investigation of the correlation between whisker roots and coating microstructures by using a planar slicing method were performed for the aforementioned two samples.

The results of these examinations clarified the mechanisms of formation and suppression of whiskers grown from the Sn–Cu coating, and we established a countermeasure against the spontaneous whisker formation through the selection of the copper lead-frame material.

ACKNOWLEDGMENTS

The JEITA's work described in Section 4.3 was mainly supported by the Organization for Small & Medium Enterprises and Regional Innovation, Japan. The authors would like to thank Prof. Suganuma of Osaka University, Mr Katsumi Yamamoto of Techno Office Yamamoto, Mr Masato Nakamura of Hitachi, Ltd, Mr Issei Fujimura of Ishihara Chemical Co., Ltd, Mr Katsuyuki Someya of Toshiba Corporation, Mr Shigeki Sakaguchi of Panasonic Corporation, Mr Yasumasa Kasuya of Rohm Co., Ltd, and all the other members of JEITA Tin Whisker Mitigation Project and JEITA Green Package Project Group for their collaboration and discussions in these project groups.

The authors of this chapter would like to thank Dr Haruo Akahoshi, Mr Masato Nakamura, Dr Takeshi Terasaki, and Dr Tomio Iwasaki at Hitachi, Ltd, Mr Tomoaki Hashimoto, Mr Kenji Yamamoto, Mr Tomohiro Murakami, Ms Yumi Yanaru, and other colleagues at Renesas Electronics Corporation for their long-term collaboration and support. They would also like to thank Dr George T. Galyon, who was at IBM eSystems and Technology Group, Ms Erika Kurauchi and Mr Tomio Hayashi at Kobelco Research Institute, Inc., and Mr Hiroaki Matsumoto and Mr Katsuji Itou at Hitachi High-Technologies Corporation for their fruitful discussions on our work presented in Section 4.4.

REFERENCES

1. Glazunova VK, Kudryavtsev NT. An investigation of the conditions of spontaneous growth of filiform crystals on electrolytic coatings. Zh Prikl Khim 1963;36(3):543–550 [in Russian].

2. Arnold SM. Repressing the growth of tin whiskers. Plating 1966;53(1):96–99.

3. Yanada I. Electroplating of lead-free solder alloys composed of Sn–Bi and Sn–Ag. IPC Printed Circuits Expo; 1998 April 26–30; Long Beach, CA, 1998, pp S11-2-1–S11-2-7.

4. Sakaguchi S, Matsushita K. Lead-free solder plating technology for outer leads of semiconductor devices. Electron Packag Technol 1999;15(8):28–35 [in Japanese].

5. Tanaka H, Tanimoto M, Matsuda A, Uno T, Kurihara M, Shiga S. Pb-free surface-finishing on electronic components' terminals for Pb-free soldering assembly. J Electron Mater 1999;28(11):1216–1223.

6. Moon K-W, Williams ME, Johnson CE, Stafford GR, Handwerker CA, Boettinger WJ. The formation of whiskers on electroplated tin containing copper. Proceedings of the 4th Pacific Rim Conference on Advanced Materials and Processing; 2001 December 11–15, Honolulu: Japan Institute of Metals; 2001, pp 1115–1118.

7. Whitlaw K, Egli A, Toben M. Preventing whiskers in electrodeposited tin for semiconductor lead frame applications. IPC/Soldertec International Conference on Lead Free Electronics; 2003 June 11–12; Brussels, Belgium; 2003.

8. Boettinger WJ, Johnson CE, Bendersky LA, Moon K-W, Williams ME, Stafford GR. Whisker and hillock formation on Sn, Sn–Cu and Sn–Pb electrodeposits. Acta Mater 2005;53(19):5033–5050.

9. Current Tin Whiskers Theory and Mitigation Practices Guideline, JEDEC/IPC Joint Publication, JP002, March 2006.

10. Nakadaira Y, Matsuura T, Tsuriya M, Vo ND, Kangas R, Conrad J, Sundram B, Lee K-H, Arunasalam SM. Pb-free plating for peripheral/lead frame packages. Proceedings of the Second International Symposium on Environmentally Conscious Design and Inverse Manufacturing (EcoDesign 2001); 2001 December 11–15; Tokyo, Japan; 2001, pp. 213–218.

11. Fujimura I. Outer lead frame plating. J Surf Finish Soc Jpn 2009;60(4):248–252 [in Japanese].

12. Jordan M. Electroplating of lead-free alloys. Second International Conference on Lead-free Electronic Components and Assemblies; 2002 December 10–11; Taipei, Taiwan; 2002.

13. Renesas Electronics. Evaluation data of lead-free terminal finishes. Available at http:// japan.renesas.com/products/lead/specific_info/rt/plating/index.jsp. Accessed 2014 Nov 19 [in Japanese].

14. IEC Standard 60068-2-69. Environmental Testing – Part 2–69: Tests – Test Te: Solderability Testing of Electronic Components for Surface Mounting Devices (SMD) by the Wetting Balance Method, IEC, Edition 2.0, May 2007.

15. Nishimura A, Sakaguchi S. Semiconductor (LQFP). JEITA Lead-Free Accomplishment Report; 2001 June 25, 2001 [in Japanese].

16. Abe S. Lead-free trends in Rohm. Electron Packag Technol 2006;22(11):54–57 [in Japanese].

17. Snugovsky P, McMahon J, Romansky M, Snugovsky L, Perovic D, Rutter J. Microstructure and properties of Sn–Pb solder joints with Sn–Bi finished components. IPC Printed Circuits Expo, APEX and the Designers Summit; 2006 February 8–10; Anaheim, CA; 2006, pp S28-01-1–S28-01-13.

18. iNEMI Tin Whisker User Group, iNEMI Recommendations on Lead-Free Finishes for Components Used in High-Reliability Products Version 4, December 1, 2006, <http://thor .inemi.org/webdownload/projects/ese/tin_whiskers/Pb-Free_Finishes_v4.pdf>, accessed October 6, 2010.

19. Satoh H, Chiba M, Takamatsu T, Kuboi T, Omae K. Evaluation of environmental and biological impact of Pb-free solder. Proceeding of Joint International Congress and Exhibition Electronics Goes Green 2004+; September 6–8; Berlin, Germany; 2004, pp 421–426.

20. Nakamura M. Verification results of internal stress type whisker mitigation technology. Proceedings of the JEITA Forum on Whisker Mitigation Technology in Electronics Packaging; March 2, 2010, pp. 79–96 [in Japanese].

21. Nishimura A, Nakamura M. Whisker mitigation effects and their mechanisms of Sn-based alloy finishes. J Jpn Inst Electron Packag 2012;15(5):357–364 [in Japanese].

22. IEC Standard 60068-2-82. Environmental Testing – Part 2-82: Tests – Test Tx: Whisker Test Methods for Electronic and Electric Components, IEC, Edition 1.0, May 2007.

23. JEDEC Standard JESD22-A121A. Test Method for Measuring Whisker Growth on Tin and Tin Alloy Surface Finishes, JEDEC, July 2008.

24. Schroeder V, Bush P, Williams M, Vo N, Reynolds HL. Tin whisker test method development. IEEE Trans Electron Packag Manuf 2006;29(4):231–238.

25. JEDEC Standard JESD201A. Environmental Acceptance Requirements for Tin Whisker Susceptibility of Tin and Tin Alloy Surface Finishes, JEDEC, September 2008.

26. Sakaguchi S, Nishimura A. Whisker evaluation results of semiconductor lead frames. JEITA Lead-Free Accomplishment Report; 2004 June 29; 2004 [in Japanese].

27. Takamizawa M, Naka T, Hino M, Murakami K, Mitooka Y, Nakai K. Effect of lead co-deposition on the whisker growth on electrodeposited tin film. J Jpn Inst Metals 2008;72(3):229–235 [in Japanese].

28. Murakami K, Okano M, Hino M, Takamizawa M, Nakai K. Mechanism of generation and suppression of whiskers on electroplated tin and tin–lead films. J Jpn Inst Metals 2008;72(9):648–656 [in Japanese].

29. Kawanaka R, Nango S, Hasegawa T, Ohtani M. Role of lead in growth suppression and growth mechanism of tin proper whisker. J Jpn Ass for Cryst Growth 1983;10(2):148–156 [in Japanese].

30. Chason E, Jadhav N, Chan WL, Reinbold L, Kumar KS. Whisker formation in Sn and Pb–Sn coatings: role of intermetallic growth, stress evolution, and plastic deformation processes. Appl Phys Lett 2008;92:171901-1–171901-3.

31. Tu KN. Irreversible processes of spontaneous whisker growth in bimetallic Cu–Sn thin-film reactions. Phys Rev B 1994;49(3):2030–2034.

32. Moon K-W, Johnson CE, Williams ME, Kongstein O, Stafford GR, Handwerker CA, Boettinger WJ. Observed correlation of Sn oxide film to Sn whisker growth in Sn–Cu electrodeposit for Pb-free solders. J Electron Mater 2005;34(9):L31–L33.

33. Zhang W, Schwager F. Effects of lead on tin whisker elimination efforts toward lead-free and whisker-free electrodeposition of tin. J Electrochem Soc 2006;153(5):C337–C343.

34. Baated A, Hamasaki K, Kim SS, Kim K-S, Suganuma K. Whisker growth behavior of Sn and Sn alloy lead-free finishes. J Electron Mater 2011;40(11):2278–2289.

35. JEITA Standard RC-5241. Whisker Test Methods for Electronic Connectors, JEITA, September 2007 [in Japanese].

36. Hillman D, Margheim S, Straw E. The use of tin/bismuth plating for tin whisker mitigation on fabricated mechanical parts. International Symposium on Tin Whiskers; 2007 April 24–25; College Park, MD; 2007.

37. Jordan M. New developments in lead free plating. IPC/JEDEC 6th International Conference on Lead Free Electronic Components and Assemblies; 2004 August 19–20; Singapore; 2004.

38. Kimizuka R, Tokio K. Whisker-suppressed tin plating and Sn–Ag alloy plating. Kinou-zairyo (Function and Materials) 2008;28(9):32–40 [in Japanese].

39. Schetty R. Tin whisker growth and the metallurgical properties of electrodeposited tin. IPC/JEDEC International Conference on Lead-free Electronic Components and Assemblies; 2002 May 1–2; San Jose, CA; 2002.

40. Smetana J. Theory of tin whisker growth: "The End Game". IEEE Trans Electron Packag Manuf 2007;30(1):11–22.

41. Yanada I, Tsujimoto M, Okada T, Murakami T. Sn–Cu alloy plating with prevention effect of Sn whisker. 12th Microelectronics Symposium (MES 2002); October 8–9; Osaka, Japan; 2002 [in Japanese].

42. Sheng GTT, Hu CF, Choi WJ, Tu KN, Bong YY, Nguyen L. Tin whiskers studied by ion beam imaging and transmission electron microscopy. J Appl Phys 2002;92(1):64–69.

43. Pedigo AE, Handwerker CA, Blendell JE. Whiskers, hillocks, and film stress evolution in electrodeposited Sn and Sn–Cu Films. Proceedings of the 58th Electronic Components and Technology Conference; 2008 May 27–30; Lake Buena Vista, FL; 2008, pp. 1498–1504.

44. Sarobol P, Pedigo AE, Su P, Blendell JE, Handwerker CA. Defect morphology and texture in Sn, Sn–Cu, and Sn–Cu–Pb electroplated films. IEEE Trans Electron Packag Manuf 2010;33(3):159–164.

45. Kato T, Akahoshi H, Nakamura M, Hashimoto T, Nishimura A. Effect of Cu lead-frame substance on whisker initiation from electrodeposited Sn/Cu coating. Proc. of IPC/JEDEC International Conference on Lead Free Electronic Components and Assemblies "RoHS Compliance and Beyond"; 2006 December; Boston, MA. Bannockburn, IL: IPC-Association Connecting Electronics Industries, 2006.

46. Kato T, Akahoshi H, Nakamura M, Hashimoto T, Nishimura A. Suppression of whisker initiation from SnCu electrodeposited coating through selection of lead frame material and it's mechanism. Proceedings of the 21st Annual Meeting of Japan Institute of Electronics Packaging (JIEP); 2007; Tokyo. Tokyo: JIEP, 2007, pp. 151–152 [in Japanese].

47. Kato T, Akahoshi H, Nakamura M, Terasaki K, Iwasaki T, Hashimoto T, Nishimura A. Correlation between whisker initiation and compressive stress in electrodeposited SnCu coating on Cu lead frame. CD Proceedings of 4th iNEMI Workshop on Tin Whiskers; 2007 May; Reno, Nevada. iNEMI, ECTC and CPMT, 2007.

48. Kato T, Akahoshi H, Nakamura M, Hashimoto T, Nishimura A. Effects of minor elements in Cu lead frame on whisker initiation from electrodeposited Sn/Cu coating. IEEE Trans Electron Packag Manuf 2007;30(4):258–269.

49. Kato T, Akahoshi H, Nakamura M, Terasaki K, Iwasaki T, Hashimoto T, Nishimura A. Whisker initiation behavior from electrodeposited Sn/Cu coating on Cu lead frame, Materials Science Forum, Vols. 561–565, pp. 1685–1688, 2007 (Trans Tech Publications, Switzerland, 2007). Also presented in PRICM-6 (The Sixth Pacific Rim International Conference on Advanced Materials and Processing), Jeju Island, Korea, November, 2007.

50. Kato T, Akahoshi H, Nakamura M, Hashimoto T, Nishimura A. Initiation and suppression mechanisms of whiskers from SnCu coating electrodeposited on lead frames. Proceedings of the 22nd Annual Meeting of Japan Institute of Electronics Packaging (JIEP); 2008; Tokyo. Tokyo: JIEP; 2008. pp 121–122 [in Japanese].

51. Kato T, Akahoshi H, Terasaki K, Iwasaki T, Nakamura M, Hashimoto T, Nishimura A. Spontaneous initiation and suppression mechanisms of whiskers from SnCu coating electrodeposited on lead frames at room temperature. Proceedings of the 23rd Annual Meeting of Japan Institute of Electronics Packaging (JIEP); 2009; Yokohama. Tokyo: JIEP: 2009. pp 161–164 [in Japanese].

52. Kato T, Akahoshi H, Nakamura M, Terasaki K, Iwasaki T, Hashimoto T, Nishimura A. Correlation between whisker initiation and compressive stress in electrodeposited tin–copper coating on copper lead frames. IEEE Trans Electron Packag Manuf 2010;33(3):165–176.

53. Teppo O. A thermodynamic analysis of the binary alloy systems Cu–Cr, Cu–Nb and Cu–V. Comput Coupling Phase Diagrams Thermochem 1990;14:125–137.

54. Peplinski B, Schulz G, Schultze D, Schierhorn E, Zakel E. X-ray powder diffraction investigation of ternary Au–Cu–Sn alloys. Mater Sci Forum 1994;166–169(pt. 2):443–448.

55. Needleman A, Rice JR. Plastic creep flow effects in the diffusive cavitation of grain boundaries. Acta Metall Overview Paper 1980;28(9):1315–1332.

56. Herring C. Diffusional viscosity of a polycrystalline solid. J Appl Phys 1950;21:437–444.

57. Tu KN. Interdiffusion and reaction in bimetallic Cu–Sn thin films. Acta Metall 1973;21(4):347–354.

58. Lee B-Z, Lee DN. Spontaneous growth mechanism of tin whiskers. Acta Mater 1998;46(10):3701–3714.

59. Oberndorff PJTL, Dittes M, Petit L. Intermetallic formation in relation to tin whiskers. In: Proceedings IPC/Soldertec International Conference Lead Free Electron; 2003 June: Brussels, Belgium; 2003.

60. Oberndorff P, Dittes M, and Crema P. Whisker testing: reality or fiction? In: Proceedings IPC/Soldertec 2nd International Conference Lead Free Electron, 2004 June; Amsterdam, The Netherlands; 2004.

61. Sakamoto I. JEITA whisker testing method statues. In: Proceedings IPC/Soldertec International Conference Lead Free Electron; 2003 June; Brussels, Belgium; 2003.

62. Sakamoto I. Whisker test methods of JEITA whisker growth mechanism for test methods. IEEE Trans Electron Packag Manuf 2005;28(1):10–16.

63. Sobiech M, Welzel U, Mittemeijer EJ, Hügel W, Seekamp A. Driving force for Sn whisker growth in the system Cu–Sn. Appl Phys Lett 2008;93:011906.

64. Iwasaki T. Application of molecular-dynamics simulation to interface stabilization in thin-film devices. Int J Jpn Soc Mech Eng, Series B 2004;47(3):470–476.

65. Hayashi T. Analytical techniques, which support nano-working -specimen working/observation techniques-. Surf Finish Soc Jpn 2005;56(7):385–392 [in Japanese].

66. Hirano S, Nakamoto T, Matsunaga S, Ueno K, Hayashi T. Crystallographic orientation analysis for generating mechanism of Sn whisker. In: Proceedings Mate 2005 Microjoining and Assembly Technology in Electron; Yokohama, Japan, 2005 [in Japanese].

5

CHARACTERIZATION TECHNIQUES FOR FILM CHARACTERISTICS

TAKAHIKO KATO

Research & Development Group, Hitachi, Ltd., Tokyo, Japan; Center for Advanced Research of Energy and Materials, Hokkaido University, Sapporo, Japan

YUKIKO MIZUGUCHI

Advanced Materials Laboratory, Sony Corporation, Atsugi, Kanagawa, Japan

5.1 INTRODUCTION

Characterization methods using transmission electron microscopy (TEM), scanning electron microscopy (SEM), and electron backscatter diffraction (EBSD) for film, one of the most powerful analytical means of determining local information, can be used to easily clarify the essential characteristics of tin and tin-based alloy finishes and thus deepen our understanding of whisker formation mechanisms. This chapter shows interesting data regarding whisker phenomena that are characterized by the most recent TEM, SEM, and EBSD techniques.

5.2 TEM (TAKAHIKO KATO)

The principle of TEM and scanning transmission electron microscopy (STEM) has been described in various text books and descriptions [1–8], and fundamentals have not been changed even to the present day. The electron beam emission gun of TEM/STEM has been modified from thermionic emission (TE) type to Schottky

Mitigating Tin Whisker Risks: Theory and Practice, First Edition.
Edited by Takahiko Kato, Carol A. Handwerker, and Jasbir Bath.
© 2016 John Wiley & Sons, Inc. Published 2016 by John Wiley & Sons, Inc.
Companion website: www.wiley.com/go/Kato/TinWhiskerRisks

emission (SE) or field emission (FE) type and thus provides high brightness, high-energy resolution, long emitter lifetime, high contrast, and high resolution. To provide high resolution, a Cs (spherical aberration coefficient) corrector has also been developed to optimize the resolving power.

There have been more requirements for energy dispersive X-ray spectroscopy (EDX) analysis; it now needs to provide high-sensitive, subnano analysis, be easy to operate, give little or no damage to the specimen, and be compatible with real-time mapping. Furthermore, high-energy resolution (e.g., <0.5 eV) and chemical analysis function has developed as a branch of electron energy-loss spectroscopy (EELS).

Another advanced analytical approach, such as nanodiffraction (nanobeam diffraction or microdiffraction), spatially resolved EELS, *in situ* observation, electron holography function, and special holder linkage have also been optionally equipped. The currently used TEM/STEM instruments, HD-2700 (Hitachi, Japan), JEM-ARM200F (JEOL, Japan), Titan G2 60-300, and Tecnai Osiris (FEI, USA), and 200Cs-TEM/STEM (Zeiss Libra, Germany) are categorized on the basis of the aforementioned requirements.

We will now introduce how the data was interpreted. Current advances were focused on the analytical functionality of TEM, STEM, and related techniques including the focused ion beam (FIB) process, EDX, and interestingly a new holder function, which are thought to be useful for studying the mechanism and mitigation of whisker formation. Concrete examples of whisker analysis that use these techniques are also described.

The typical indexed electron diffraction patterns (EDPs) for materials that are usually observed in tin and tin-based alloy finishes and at the interface between the finishes and the substrates are recorded. The primary materials with the indexed EDPs are tin (Sn), tin oxides (SnO and SnO_2), and intermetallic compounds (Cu_6Sn_5, Cu_3Sn, Ni_3Sn_4, and $FeSn_2$).

5.2.1 TEM Analysis

Figure 5.1 (T. Kato et al., unpublished data) shows a typical example of a field emission transmission electron microscopic (FE-TEM) observation result for a whisker formed during the thermal cycle test from electrodeposited tin–copper coating on an Alloy 42 substrate using a Hitachi HF-2000 system. The TEM thin-foil specimen was obtained by the FIB process after protection of the whisker by three-layered vacuum vapor deposition films consisting of carbon, platinum, and tungsten. It is clear from the FE-TEM bright-field image that as the whisker grows, it pushes up a protective resin on the coating.

Furthermore, the whisker consists of a single crystal because the whisker has the same diffraction patterns, and this indicates a β-tin structure from whisker root to whisker tip. These patterns were obtained by using a conventional diffraction method with the selected area from regions 1 to 5 shown by circles and by using a nanodiffraction method with a 1 nm diameter electron beam from region 6 as shown at the point-marking position. A comparison of diffraction patterns from tin–copper coating grains clearly shows that the crystalline orientation of the whisker is entirely different

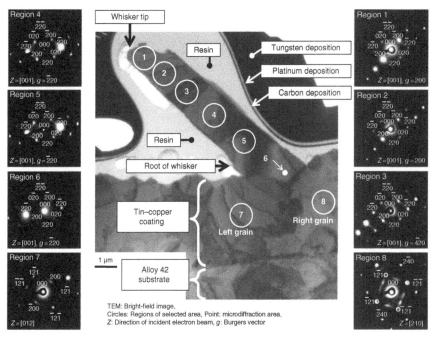

Figure 5.1 Typical field emission transmission electron microscopy (FE-TEM) analysis at cross section of whisker formed from electrodeposited tin–copper coating on Alloy 42 substrate during the thermal cycle test. Diffraction patterns from whisker and coating indicate β-tin structure.

from that of coating grains located just below the whisker root. Figure 5.2 (T. Kato et al., unpublished data) shows the FE-TEM observation result at the cross section of the whisker formed from electrodeposited tin–copper coating on a copper substrate during a corrosion test under 93% relative humidity (RH) at 333 K (60°C) for 2000 h.

The specimen was pre-heat-treated at 423 K (150°C) for 1 h before the corrosion test. The left whisker was grown from the noncorroded tin–copper coating side, which is located very near the coating-oxide interface. It was clear from the photograph that the thickness of the coating of the corroded part (right side) was thicker than that of the noncorroded coating (left part). This result easily helps us to understand that the whisker formation under corrosion atmosphere occurred as a consequence of compressive stress generation in the coating near the coating-oxide interface.

It was of further interest that the whisker located at the corroded coating (right side) is itself also corroded from the whisker root during the corrosion progress of the coating film from the right to the left side of the coating. Figure 5.3 (T. Kato et al., unpublished data) indicates nano-diffraction patterns (obtained by using a 1-nm-diameter electron beam) from the indicated points of (1)–(6) in Figure 5.2. From the nano-diffraction patterns of (1)–(3), we clarified that the oxide of the coating near the noncorroded coating oxide interface consists of a mixture of SnO and SnO_2 crystals.

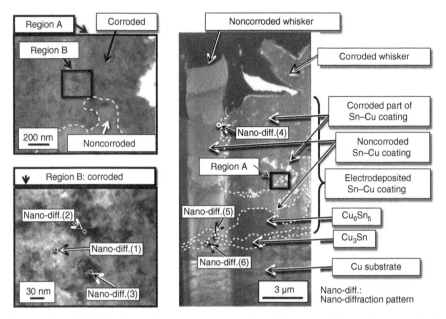

Figure 5.2 FE-TEM observation at cross section of whisker formed from electrodeposited tin–copper coating on copper substrate during a corrosion test under 93% relative humidity at 333 K. Nano-diffraction patterns from indicated points (1)–(6) are shown in Figure 5.3.

Analysis of other nano-diffraction patterns can be used to characterize not only broad areas such as the tin–copper coating with a β-Sn structure (nano-diffraction (4)) but also extremely limited analytical areas of Cu_6Sn_5 or Cu_3Sn (nano-diffractions (5) and (6)). As mentioned earlier, reliable microstructural and nanolevel crystallographic characterizations of whiskers and their surrounding regions can be obtained by using current TEM technologies.

5.2.2 STEM and EDX Analysis

Figure 5.4 (T. Kato et al., unpublished data) shows (a) STEM (dark field), (b) STEM (bright field), and (c) SEM images obtained by the FE-STEM instrument for the same thin-foil specimen with whiskers. These images can be switched without any adjustment of the electron beam in this FE-STEM series, as well as the Hitachi HD-2000 and HD-2700 series. In the STEM dark-field (Z-contrast) image, the white contrast intensity W-Int. is a function of mass number Z, density ρ, and proportionality constant α, as expressed by W-Int. $= \alpha(Z^2 \times \rho)$ [9]. For example, the shape and location of iron particles in the copper lead frame are clearly identified as black contrasts because of their relatively lower W-Int. as shown in Figure 5.5 [9]. The iron particles exist in the Cu_7Sn_5 intermetallic compound layer [10, 11] as well. This mechanism is described in Section 4.4.5 on "Kirkendall markers."

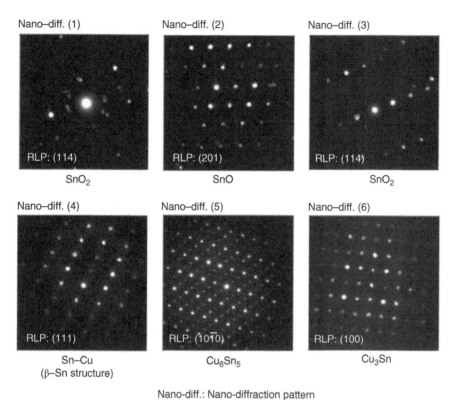

Nano–diff. (1)   Nano–diff. (2)   Nano–diff. (3)

RLP: (114)   RLP: (201)   RLP: (114)

SnO_2 SnO SnO_2

Nano–diff. (4)   Nano–diff. (5)   Nano–diff. (6)

RLP: (111)   RLP: ($10\bar{1}0$)   RLP: (100)

Sn–Cu Cu_6Sn_5 Cu_3Sn
(β–Sn structure)

Nano-diff.: Nano-diffraction pattern
RLP: Reciprocal lattice plane

Figure 5.3 FE-TEM nano-diffraction patterns and identified phases. Nano-diffraction patterns are obtained from indicated points (1)–(6) shown in Figure 5.2.

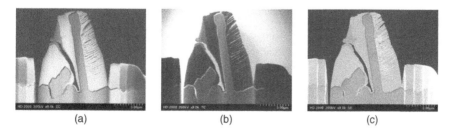

(a)   (b)   (c)

Figure 5.4 (a) STEM (dark field), (b) STEM (bright field), and (c) SEM images obtained by FE-STEM instrument, Hitachi HD-2000, for same thin-foil specimen with whiskers. Cross-sectional specimen was made by FIB process from electrodeposited tin–copper coating on Alloy 42 substrate after thermal cycle test.

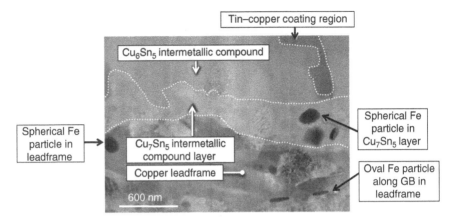

Figure 5.5   FE-STEM (HD-2000) dark-field (Z-contrast) image for cross-sectional sample of electrodeposited tin–copper (Sn–2 mass%Cu) coating on copper lead frame (CDA number: C19400). Source: Reproduced by permission of the Institute of Electrical and Electronics Engineers [9].

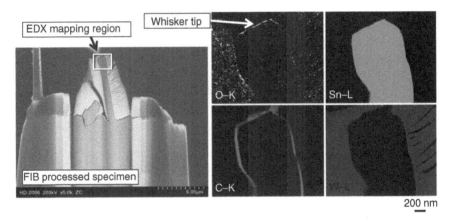

Figure 5.6   FE-STEM (dark field) image and result of EDX mapping for whisker top region. The cross-sectional specimen was made by FIB process from electrodeposited tin–copper coating on Alloy 42 substrate after thermal cycle test.

Another utility is the EDX mapping function. Figure 5.6 (T. Kato et al., unpublished data) shows EDX mapping of oxygen, tin, carbon, and tungsten at the very top of the whisker, which is the same whisker as that in Figure 5.4. The result shows that thick oxide film is formed only on the whisker tip, and it mainly consists of tin. Before FIB processing of the specimen, the surface of the whisker was protected by carbon and tungsten depositions as indicated in Figure 5.6.

Figure 5.7 (T. Kato et al., unpublished data) indicates another FE-STEM/EDX mapping result at the cross section of electrodeposited tin coating on the copper

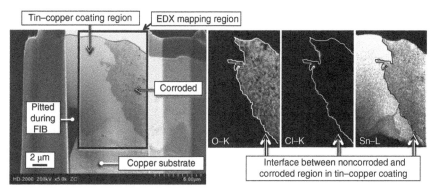

Figure 5.7 FE-STEM/EDX mapping result at cross section of electrodeposited tin–copper coating on copper substrate sample after corrosion test in 87% relative humidity at 333 K for 4000 h. Sample was heat-treated at 423 K for 1 h before corrosion test.

substrate sample after a corrosion test in 87% RH at 333 K (60°C) for 4000 h. Chlorine atoms exist along the tin coating–tin oxide interface. This means that chlorine is one of the driving elements for oxidation of tin coating in this system under the test condition. The EDX analysis for minor elements confirmed that the elemental peak in the X-ray spectrum (intensity vs X-ray energy) was authentic. This was because the EDX mapping system can automatically indicate a para-signal from the background of different element(s).

Figure 5.8 shows dark-field STEM (DF-STEM) images of Au (100) single crystal by using the original STEM and Cs-corrected STEM. As spherical aberration is almost negligible in this case, higher resolution has been obtained (point resolution has been increased from 0.23 to 0.14 nm). The Cs-corrected STEM is also appropriate for precisely observing boundary structures such as grain boundaries and interfaces of different materials.

5.2.3 FIB Process

Figure 5.9 (T. Kato et al., unpublished data) shows an example of the FIB process for sampling a specimen with whiskers. The sample is a corrosion-tested lead consisting of experimentally electrodeposited tin-bismuth coating on a copper lead frame. The corrosion test condition was 93% RH at 333 K (60°C) for 2000 h. After identifying the whiskers by optical microscope, or scanning ion microscope (SIM) ((a)–(c) in the figure) on the lead surface, target whiskers have been protected by depositions (d), and then the TEM sample with whiskers was collected by using the FIB process shown in (e). The TEM thin-foil specimen (f) was obtained by continuous FIB thinning. Surface roughness after FIB has been currently improved by a gentle-milling technique and other techniques to take higher quality TEM/STEM images (including lattice and structural images).

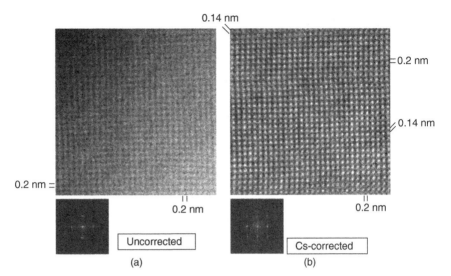

Figure 5.8 Dark field-STEM (DF-STEM) images of Au(100) single crystal by using the original STEM (a) and Cs-corrected STEM (b) (Hitachi, HD-2700). Lower images are power spectra of each DF-STEM image. Convergence semiangle of electron probe is 8- and 25-mrad for original and corrected STEM, respectively. Source: Courtesy of Hitachi High-Technologies Corporation.

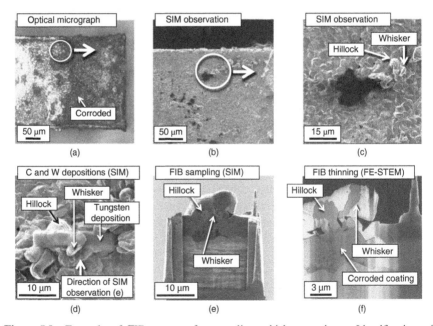

Figure 5.9 Example of FIB process for sampling whisker specimen. Identification of whiskers from normal direction to surface of lead specimen by optical microscope (a) and scanning ion microscope (SIM) (b)–(d). Cross-sectional micrographs (e) progress of FIB processing and (f) FIB completion for same hillock and whisker with (c) and (d).

5.2.4 Holder Techniques

Many special kinds of sample holder have been frequently developed besides conventional ones and include the following functionality: an environmental atmosphere for influencing a specimen, elevated temperature, *in situ* observation, a 3D analytical holder, compatibility with revolver, and so forth. Even conventional holders provide new functions for various kinds of users. For example, holders made of a wide variety of substances, that is, titanium, copper, or aluminum holders, can be used to avoid system contaminations or contamination from mount meshes of specimens in EDX analysis. Holders that do not require specimens to be remounted can be used with different TEM/STEM instruments and are popular. An air-protected holder for using lithium-contained specimens to study lithium ion battery materials is also widely used.

There is a 3D analytical holder with 360° rotation for use in future whisker investigations. Figure 5.10 indicates a micropillar sampling process for pinpoint 3D analysis. The sampling can make micropillar specimen that when used with the 360° rotation holder (3D analytical holder) [12] (shown in Fig. 5.11) will enable

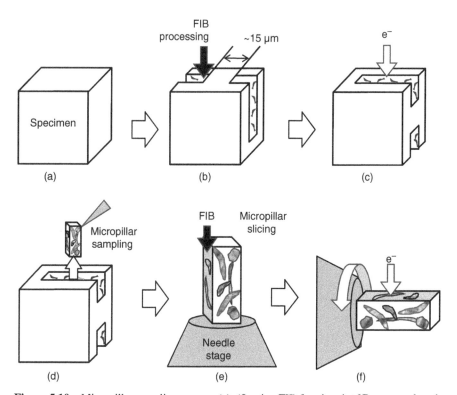

Figure 5.10 Micropillar sampling process (a)–(f) using FIB for pinpoint 3D structural analysis. Source: Courtesy of Hitachi High-Technologies Corporation.

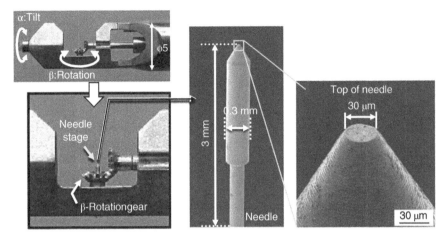

Figure 5.11　Appearance of 360° rotation holder and needle for pillar sample analysis. Source: Courtesy of Hitachi High-Technologies Corporation.

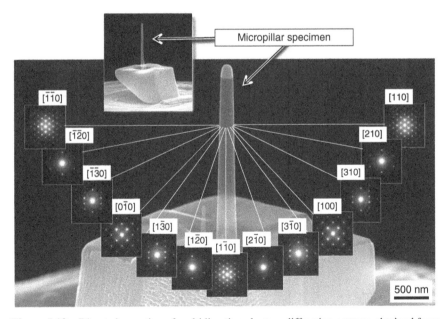

Figure 5.12　Direct observation of multidirection electron diffraction patterns obtained from micropillar specimen made by Si single crystal using 3D analytical holder, which allows 360° rotation of specimen. Source: Courtesy of Hitachi High-Technologies Corporation.

direct observation of multidirection EDPs obtained from the micropillar specimen. Figure 5.12 shows an example of direct observation of multidirection EDPs obtained from the micropillar specimen made by a Si single crystal using the 3D analytical holder. As such, this special holder and micropillar sampling process are thought useful for filament-like whisker investigations.

5.2.5 Indexed Diffraction Patterns in Relation to Whisker Studies

Populated reciprocal lattice planes, which correspond to the EDPs that often appear in lead-free tin and tin-based materials having whisker problems, are recorded here. Readers may refer to them when indexing the diffraction patterns obtained during investigating whisker mechanisms and mitigations. Tin and tin oxides (SnO and SnO_2), and intermetallic compounds Cu_6Sn_5, Cu_3Sn, Ni_3Sn_4, and $FeSn_2$, which frequently appear as the object materials near the interface between the tin or tin-based alloy finishes and the substrates of copper or Alloy 42, are mainly included. These populated reciprocal lattice planes are shown in Figures 5.13–5.21 (T. Kato et al., unpublished data).

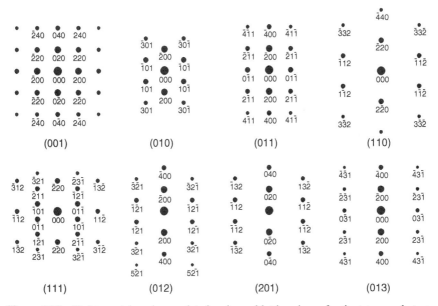

Figure 5.13 Eight most densely populated reciprocal lattice planes for the tetragonal structure of β-Sn.

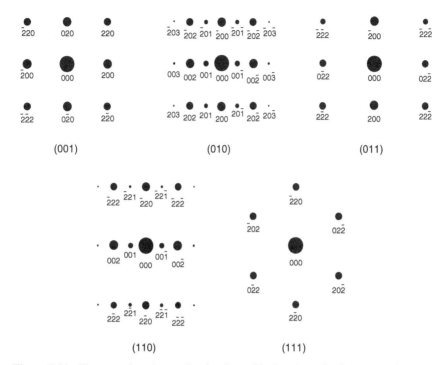

Figure 5.14 Five most densely populated reciprocal lattice planes for the tetragonal structure (equilibrium phase) of SnO.

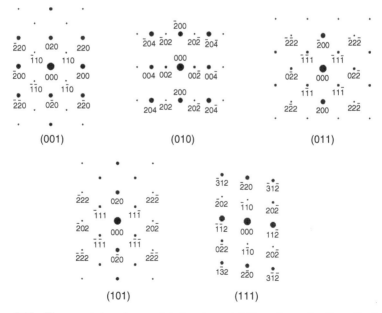

Figure 5.15 Five most densely populated reciprocal lattice planes for the orthorhombic structure (metastable phase) of SnO.

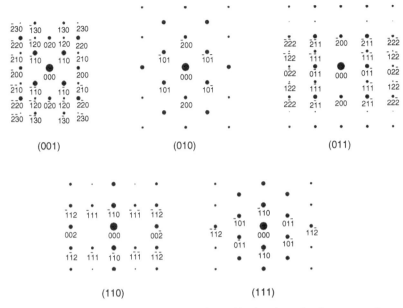

Figure 5.16 Five most densely populated reciprocal lattice planes for the tetragonal structure of SnO_2.

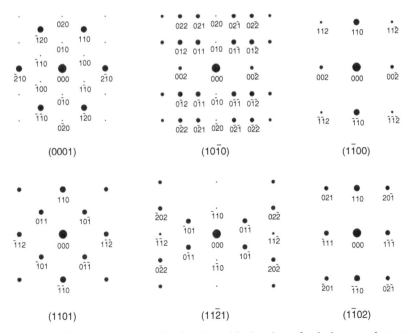

Figure 5.17 Six most densely populated reciprocal lattice planes for the hexagonal structure of Cu_6Sn_5.

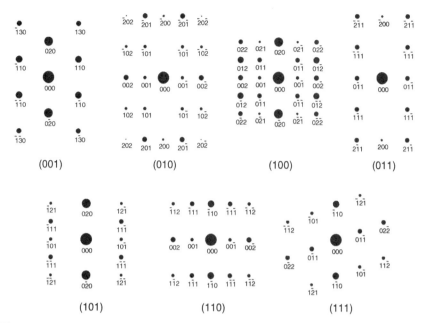

Figure 5.18   Seven most densely populated reciprocal lattice planes for the orthorhombic structure of Cu_3Sn (low temperature phase).

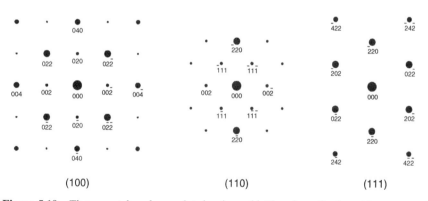

Figure 5.19   Three most densely populated reciprocal lattice planes for the cubic structure of Cu_3Sn (high temperature phase).

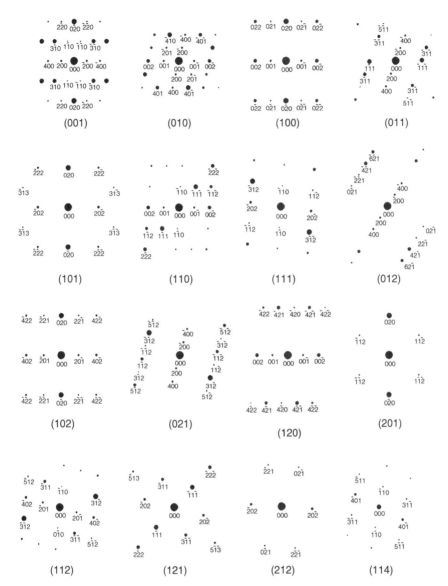

Figure 5.20 Sixteen most densely populated reciprocal lattice planes for the monoclinic structure of Ni_3Sn_4.

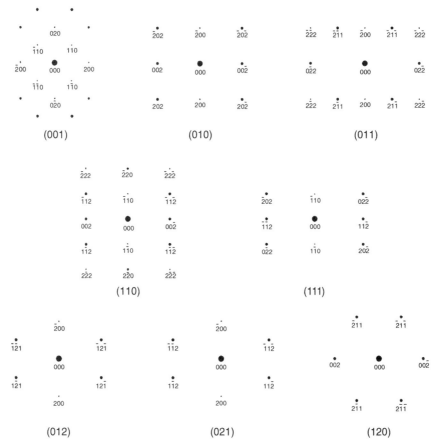

Figure 5.21 Eight most densely populated reciprocal lattice planes for the tetragonal structure of FeSn₂.

5.3 SEM (YUKIKO MIZUGUCHI)

Typical Sn (tin) whiskers are 1–5 μm in diameter and range from submicrometers to several thousand micrometers in length. These whiskers can have a straight (filamentary), kinked, curved, or even winding shape. To observe the topography of objects of this size, the scanning electron microscope (SEM) is indispensable.

The SEM is a type of electron microscope that can form images of sample surface topography and composition. When a focused electron beam is scanned across a solid sample, the electrons interact with the sample surface, producing secondary electrons, backscattered electrons, and X-rays.

Secondary electrons provide images from the topographic contrast of the sample surface. The practical observation range of SEM is from submicrometers to several thousand micrometers. Thus, SEM (especially secondary electron imaging) is well suited for observation of whisker topography.

The principles of SEM are fully described in comprehensive texts [13, 14]. This section introduces several examples of SEM image use in tin whisker analysis: observation of tin whisker shape, measurement of tin whisker length and kink angle, and confirmation that tin whiskers do indeed generate from the base metal.

5.3.1 Observation of Tin Whisker Topography

Figure 5.22 shows a SEM image of tin whiskers formed under an externally applied mechanical stress (hereafter, this whisker type is referred to as "mechanically induced whiskers") on a connector pin [15]. The sample was observed along the z-direction (the definition of the sample coordinate system is shown in the figure). Many filamentary whiskers with a maximum length exceeding 100 μm can be clearly observed on the connector pin. Clearly, a topographic image of the whiskers can be acquired by SEM.

Figure 5.23 shows an enlarged SEM image of some of the whiskers shown in Figure 5.22. Typically, tin whiskers have a straight (filamentary) shape, and as these filamentary whiskers grow, they change their growth direction by kinks. Basically, filamentary whiskers have continuous SEM contrast aligned with the whisker growth directions, as shown in (a). However, they often contain characteristic SEM contrast, as shown in (b), (c), and (d), which are not continuously aligned with the whisker growth directions indicated by the white arrows (they are instead shaped as bamboo joints (b), and kinks (c) and (d)). These discontinuous shapes can be attributed to the formation of lattice defects such as dislocations, twinning, or grain boundaries during whisker growth [15].

Round whiskers can be formed in certain peculiar environments (such as during thermal cycling). Such whiskers can be generated by thermal expansion mismatch

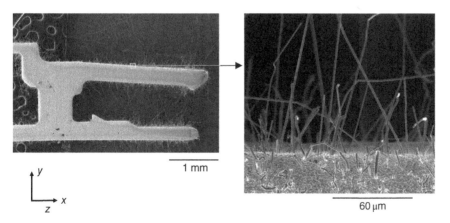

Figure 5.22 Mechanically induced whiskers on a connector pin. Source: Reproduced by permission of the Institute of Electronics, Information and Communication Engineers.

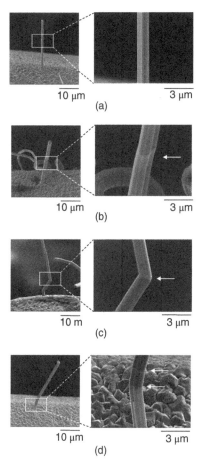

Figure 5.23   Enlarged SEM image of some whiskers in Figure 5.22. Source: Reproduced by permission of the Institute of Electronics, Information and Communication Engineers.

between tin plating and a substrate with a low thermal expansion coefficient, such as a ceramic or Alloy 42, during thermal cycling (hereafter, whiskers formed by thermal cycling are referred to as "thermal cycling whiskers"). Suganuma et al. examined the growth mechanism of thermal cycling whiskers using microstructural observation by SEM and TEM (see Section 5.2 for transmission electron microscope (TEM) techniques) [16].

Figure 5.24 shows SEM images of a typical thermal cycle whisker with an enlarged image of the side view. Very fine striation rings indicated by arrows are clearly observable from the bottom to the top of the whisker. These striation rings have intervals of about 100–200 nm, and their analysis revealed that each striation corresponds to one thermal cycle. In addition to the striation rings, deep grooves were observed at the bottom of the whisker, as indicated by the thick arrow. The formation of these

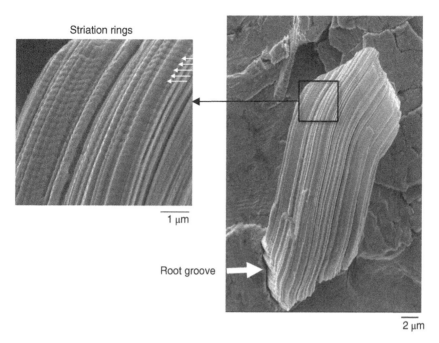

Figure 5.24   SEM images of a typical thermal cycle whisker with an enlarged image of the side face. Source: Reproduced by permission of Elsevier.

whisker root grooves can be attributed to tin oxidation along the grain boundary at the tip of the whisker root groove during thermal cycling. Suganuma et al. concluded that the formation of whisker root grooves by oxidation plays an important role in the growth of winding whiskers.

SEM is an important tool for the observation of whisker topography, as shown in Figures 5.22–5.24, and can provide clues to a better understanding of whisker growth mechanisms.

5.3.2   Measuring the Kink Angle and Length of Tin Whiskers

Whiskers change their growth direction with kinks, as shown in Figures 5.22 and 5.23. This means that a given whisker can grow in any direction and potentially cause shorts in electrical parts. Measuring the length and kink angles of such whiskers is very important, and several measurement techniques have been proposed [17–22]. However, these methods often require considerable time and effort. Here, there is introduced one simple method of measuring whisker length and kink angle.

Figure 5.25 shows SEM images of a whisker taken from three different directions. Although all three SEM images show the same whisker, the kink angles (white arrows) and length appear to be different. Therefore, to measure the whisker length and kink angle accurately, operators should pay careful attention to the measurement method.

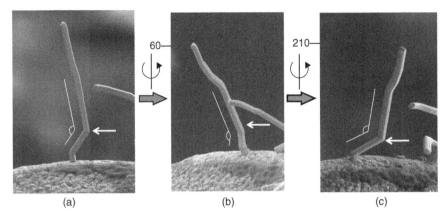

(a)   (b)   (c)

Figure 5.25   SEM images of a whisker that was observed from three different directions. (a) SEM image of a whisker (reference), (b) the whisker observed from 60° rotated direction, against (a), and (c) the whisker observed from 210° rotated direction, against (b).

Figure 5.26   A whisker with the kink angle of 45°. Source: Reproduced by permission of the Institute of Electronics, Information and Communication Engineers.

Kinks typically have certain dominant angles, such as approximately 30°, 45°, 60°, and 90° [15, 21–23]. Kink formation is related to plastic deformation, such as sliding and twinning [15, 21]. Because the sliding and twinning of tin depend on the tin crystal orientation, certain kink angles should be more common than others [24–26].

Figure 5.26 shows an example of a whisker with a kink angle of 45°. Whisker axes 1 and 2, which form the kink, define a plane. By using the SEM to observe the kink from a direction perpendicular to this plane, the correct measurement of the kink angle can be ensured.

Figure 5.27 shows an improper example of kink angle measurement. A SEM has a finite depth of focus. If a kink is not observed from a direction perpendicular to the plane containing whisker axes 1 and 2, the focus cannot be adjusted to include both

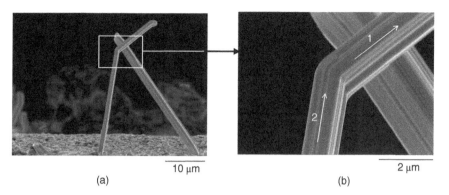

     (a)   (b)

Figure 5.27 An improper example of observation for estimating kinked angle.

axes 1 and 2. In Figure 5.27, the focus was adjusted to axis 2, but axis 1 remained out of focus. Therefore, this observation was not performed along a direction perpendicular to the plane containing both whisker axes, so it cannot provide a correct measurement of the kink angle. This phenomenon can be more clearly observed by operating at a higher magnification, as shown in Figure 5.27b. Similarly, when the focus was adjusted to axis 1, axis 2 fell out of focus (the proper high magnification depends on the whisker length, thickness, and the SEM equipment). Observing the kink from a direction perpendicular to the plane containing both whisker axes is necessary in order to correctly measure the kink angle.

As in the measurement of kink angle, accurate whisker length measurement by SEM requires placing the area of interest (such as the segment of a whisker that falls between two successive kinks) in a plane perpendicular to the direction of observation.

If it is impossible to place a whisker in a single plane, as when a whisker vents in many different directions, other methods can be used [17–22].

5.3.3 Confirmation of Tin Whiskers Generate from These Base Metals

In peculiar environments and by using peculiar materials, whiskers can be grown by the vapor–liquid–solid method, in which whiskers are grown by direct adsorption of a gas phase [27, 28]. However, all tin whiskers are believed to be generated and grown from a base metal (pure tin or a tin-based alloy). To confirm whether tin whiskers do indeed generate from these base metals, whisker growth was observed by SEM [29], and the typical result is shown in Figure 5.28. Acrylic particles 3 µm in diameter were attached to Sn–Cu plating with glue as markers, and a mechanical stress was applied to the plating. Mechanically induced whiskers have much higher growth rates than spontaneous whiskers [15, 30–33]. (Spontaneous whiskers are formed by the compressive stresses induced by the formation of Cu_6Sn_5 intermetallic compounds; see Chapter 4 for more information on spontaneous whiskers.)

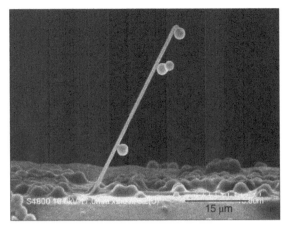

Figure 5.28   Observation of whisker with markers. Confirmation of whiskers generated from base.

In the present experiment, the whiskers were generated with markers (the particles attached to Sn–Cu plating) in a few days, indicating that the whiskers grew from the base alloy (not by direct adsorption of a gas phase). This result indicates that tin whiskers do indeed generate from these base metals.

In conclusion, several examples of the usage of SEM images for tin whisker analyses were discussed: observation of tin whisker shape, measurement of tin whisker length and kink angles, and confirmation of tin whiskers do indeed generate from these base metals. SEM is a very useful tool for the investigation of whisker formation mechanisms.

5.4   EBSD (YUKIKO MIZUGUCHI)

Sn (tin) crystals have anisotropic mechanical properties due to their anisotropic crystal structure (β-Sn structure, which is body-centered tetragonal with lattice parameters $a = b = 0.583$ nm and $c = 0.318$ nm at 25°C, where $c/a = 0.546$). Young's modulus along the c-axis is about three times greater than that along the a- and b-axes [25, 34]. Thus, tin whisker formation is affected by the crystal orientation of tin grains, so understanding the crystal orientation relationship between a whisker and its adjacent grains is important for clarifying the tin whisker formation mechanism.

This section briefly describes general crystal orientation characterization techniques and then focuses on the principles of and advances in EBSD and its use in investigating the effects of crystal orientation on tin whisker formation (EBSD can also be referred to as electron backscatter diffraction pattern (EBSP)). Finally, several EBSD-based investigations of the effects of crystal orientation on tin whisker formation are discussed.

5.4.1 Crystal Orientation Characterization Techniques

TEM and XRD (X-ray diffraction) have long been used to investigate crystal orientation in metals. In recent years, EBSD has become another popular crystal orientation characterization technique for use on metals.

EBSD was first developed by Alam et al. in 1954 [35, 36]. In the 1970s, Venables and Harland applied EBSD to analyze the crystal orientation of silicon, tungsten, and aluminum in a SEM [37]. Since the 1990s, EBSD has become more widely available and easier to use because of the rapid increase in computer processing speed.

Figure 5.29 shows the practical characterization scales of TEM, XRD, and EBSD for identifying the crystal orientation of objects.

TEM is capable of a significantly higher resolution than EBSD or XRD. TEM can detect crystal orientation even in nanometer-sized microstructures. However, because TEM methods make it difficult to analyze wide-scale (over several tens of micrometers) crystal orientation, the analytical data does not necessarily represent the overall crystal orientation of an object. On the other hand, XRD is capable of identifying wide-scale (over several hundreds of micrometers) crystal orientation, and XRD data can indicate preferred orientations of the objects (e.g., a polycrystalline sample).

However, it is difficult to use XRD to identify local crystal orientation, such as the orientation of each grain in a polycrystalline sample. For polycrystalline samples, X-ray microdiffraction using synchrotron radiation enables the identification of the orientation of each grain in a polycrystalline sample. However, to analyze crystal orientation by means of X-ray microdiffraction, a special synchrotron radiation facility is needed, such as SPring-8 (in Japan), the Advanced Photon Source (in America), or the European Synchrotron Radiation Facility (in France).

EBSD can be used to identify the crystal orientation of each grain in a polycrystalline sample, with an applicable grain size ranging from submicrometers to several thousand micrometers. In addition, EBSD can detect the preferred orientations of objects over a wide area (from submicrometers to several thousand micrometers). Because EBSD can analyze both the local orientation of the whiskers and the preferred orientation of the whisker's surroundings, EBSD is an indispensable technique for clarifying the whisker formation mechanism.

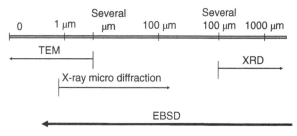

Figure 5.29 Practical characterization scales of TEM, XRD, and EBSD for identification of the crystal orientation of objects.

5.4.2 Principles of EBSD

Figure 5.30 shows a schematic illustration of an EBSD measurement. EBSD measurements are conducted in an SEM (Section 5.3) equipped with an EBSD detector, which contains a phosphor screen and a CCD camera. A polished or flat sample is placed in the SEM chamber and tilted to a high angle (typically 70°). When the electron beam is scanned across the sample surface, it diffracts at specific crystal lattice planes. The diffracted beam produces a pattern composed of characteristic bands, which are called EBSP. These patterns are then imaged by the phosphor screen onto the CCD camera.

Figure 5.31 shows an EBSP of β-Sn. The bands in the pattern are referred to as Kikuchi bands. The appearance, for example, the width of the bands and the angles between the bands, of the pattern is directly related to the crystal lattice structure and crystal orientation at the point where the beam meets the sample. Thus, the EBSD technique allows not only microstructural phase and crystal orientation information of each sampled point but also various information related to the crystal orientation of any area of interest. For example, the distribution of crystalline phases (phase map), the preferred orientation (pole figure and inverse pole figure), the distribution of crystal orientation parallel to certain sample directions (inverse pole figure map (IPF map)), and the grain size can be determined by combining the phase and crystal orientation information of any area of interest.

The EBSP is, in principle, very similar to a Kikuchi pattern obtained via TEM.

5.4.3 Determining the Crystal Orientation of a Whisker and Its Adjacent Grains

This section discusses several examples of crystal orientation analysis of a whisker and its adjacent grains, using both spontaneous whiskers and whiskers formed by

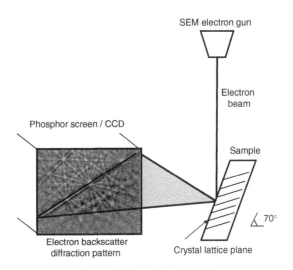

Figure 5.30 Schematic illustration of EBSD measurements.

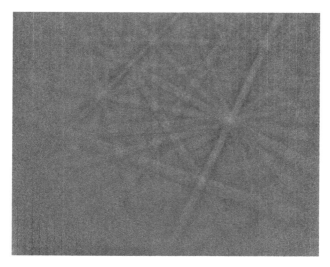

Figure 5.31   An example of electron backscattered diffraction pattern of β-Sn.

an externally applied mechanical stress (hereafter, whiskers formed by an externally applied mechanical stress are referred to as "mechanically induced whiskers").

Mizuguchi et al. performed cross-sectional EBSD analysis on mechanically induced whiskers formed on an Sn–Cu plating [30]. Figure 5.32a shows an IPF map along the sample z-direction (normal to the plating surface; z_{sc}), in which the cross-sectioned whisker and its vicinity are shown. The colors in the map indicate the measured crystal orientations according to the color key (standard triangle of stereogram). For example, the purple colored grain is $\langle 111 \rangle$ oriented in the sample z-direction. This figure clearly shows that the whisker formed on columnar grains as a single crystal. The whisker grew almost parallel to the sample z-direction (indicated by the yellow arrow).

Figure 5.32b shows a $\langle 111 \rangle$ pole figure of tin (hereafter, denoted as $\langle 111 \rangle_{Sn}$) in the sample z-direction, in which the whisker and the grains directly beneath the whisker (grains 1 and 2) in Figure 5.32a are shown. The actual crystal orientation of the whisker is indicated by a tetragonal prism. The red, green, and blue lines in the tetragonal prisms represent the a, b, and c axes, respectively. The crystal direction aligned with the whisker axis (the aforementioned yellow arrow in Fig. 5.32a) was determined to be [11$\bar{1}$]. In this way, the crystal direction aligned with the whisker axis can be estimated using an IPF map and a pole figure.

Figure 5.32c shows a $\langle 100 \rangle$ pole figure of tin (hereafter, denoted as $\langle 100 \rangle_{Sn}$) in the sample z-direction, in which the whisker and the grains directly beneath the whisker (grains 1 and 2) in Figure 5.32a are shown. The actual crystal orientations of the tin grains of interest are indicated by tetragonal prisms in the same way as in Figure 5.32b. The analysis revealed that the whisker and one of the grains directly beneath it (grain 1) were misoriented by about 62° about the $\langle 100 \rangle$ axis; this is a well-known characteristic of tin mechanical twins with $\{301\}$ boundaries

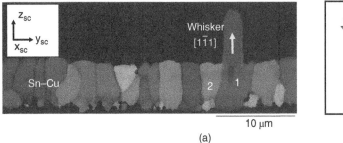
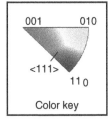

(a)

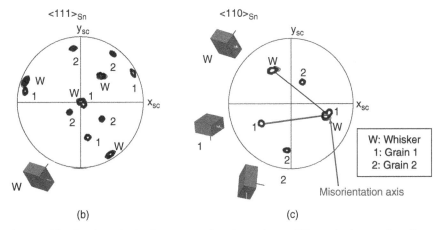

(b)   (c)

Figure 5.32 EBSD analysis of a cross-sectioned sample: (a) IPF map in the sample z direction, in which a whisker root and its vicinity are shown. (b) $\langle 111 \rangle$ and (c) $\langle 100 \rangle$ pole figures of tin in the sample z direction, in which the whisker and its adjacent grains 1 and 2 in (a) are shown. Source: Reproduced by permission of Springer Science and Business Media. Please visit www.wiley.com/go/Kato/TinWhiskerRisks to access the color version of this figure.

[25, 34, 38, 39]. Similar analysis revealed that mechanically induced whiskers formed on columnar grains had {301} twinning relations to the grains directly beneath them.

Furthermore, many whisker segments with a bamboo-joint-like appearance or kinks, as shown in Figure 5.23b and c, have twin structures: Figure 5.33 shows IPF maps along the sample z-direction (normal to the plating surface; z_{sc}), in which the cross-sectioned whisker and its vicinity are shown. The whisker segments with a bamboo-joint-like appearance or kinks have twin structures (see Ref. [15] for more details).

These studies demonstrate that plastic deformation, especially twin formation, can play a critical role in the formation and growth of mechanically induced whiskers.

Kato et al. performed EBSD analysis of horizontal cross sections of spontaneous whiskers formed on an Sn–Cu coating (see Figs. 4.51 and 4.52 in Section 4.4) [11]. Their analysis showed that whisker roots were located on the grain boundaries of the coating. Some whisker roots had a bicrystal structure, while the others had a

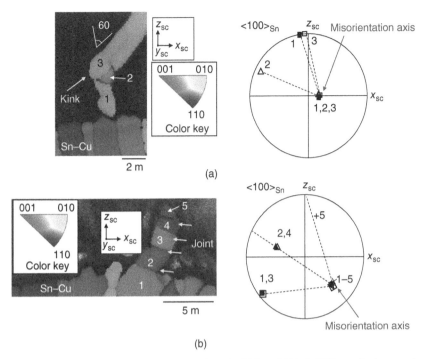

(a)

(b)

Figure 5.33 EBSD analysis of a cross-sectioned whisker: (a) Left: IPF map in the sample z direction, in which a kinked whisker and its vicinity are shown. Right: $\langle 100 \rangle$ pole figures of tin, in which the whisker segments 1, 2, and 3 in the left map are shown. Segments 1 and 2, and 2 and 3, respectively, have twin structures. (b) Left: IPF map in the sample z direction, in which a whisker with a bamboo-joint-like appearance and its vicinity are shown. Right: $\langle 100 \rangle$ pole figures of tin, in which the whisker segments 1–5 in the left map are shown. Segments 1 and 2, 2 and 3, 3 and 4, and 4 and 5, respectively, have twin structures. Source: Reproduced by permission of the Institute of Electronics, Information and Communication Engineers. Please visit www.wiley.com/go/Kato/TinWhiskerRisks to access the color version of this figure.

single-crystal structure. Furthermore, the bicrystal structure was the cause of kink formation. In addition, the whisker roots had very similar crystal orientations to the grains directly beneath the whiskers.

The EBSD results of Mizuguchi et al. and Kato et al. suggest that mechanically induced whiskers and spontaneous whiskers have different formation mechanisms, despite having very similar appearances in that they both form at or near the junctions of underlying columnar grains.

As stated earlier, the crystal orientation relationship between a whisker and its adjacent grains can be determined by analysis of an IPF map and/or a pole figure.

5.4.4 Determining the Preferred Orientation of a Whisker and Its Adjacent Grains

This section introduces several example analyses of the preferred orientation of tin on a Sn–Cu plating, which is prone to generate mechanically induced whiskers.

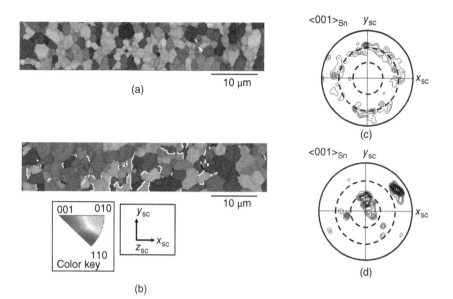

Figure 5.34   EBSD analysis for stressed and unstressed sample surface. (a) IPF map of as-deposited plating and (b) IPF map of plating to which mechanical stress had been applied. The black and white lines in (a) and (b) indicate {101} and {301} twin boundaries, respectively. (c) ⟨001⟩ pole figures showing orientations of all grains in (a). (d) ⟨001⟩ pole figures showing orientations of all grains in (b). Source: Reproduced by permission of Springer Science and Business Media. Please visit www.wiley.com/go/Kato/TinWhiskerRisks to access the color version of this figure.

Figure 5.34a and b shows IPF maps of a $262\,\mu m \times 50\,\mu m$ area in the sample z-direction, respectively, showing the surfaces of the as-deposited plating and the plating to which a mechanical stress was applied. These measurements were made in the same area on different parts of an electrical device (connector pins). The black and white lines in these maps respectively indicate boundaries misoriented by about 58° and 62° about the ⟨100⟩ axis. These relationships are well-known characteristics of mechanical twins of tin with twin boundaries of {101} and {301}, respectively [25, 34, 38, 39]. The analysis revealed that the tin grains became larger upon the application of a mechanical stress. In addition, many twins were formed. Most of the twin boundaries were {301}; the plating to which mechanical stress had been applied had 20 times more {301} twin boundaries than the as-deposited plating.

Figure 5.34c and d shows ⟨001⟩ pole figures of Sn (hereafter, denoted as $⟨001⟩_{Sn}$) in the z-direction of the plating surface shown in Figure 5.34a and b, respectively. The as-deposited plating had a preferred orientation with the ⟨001⟩ axis oriented about 60° away from the sample z-direction. After the mechanical stress was applied, these preferred orientations split into several directions due to microstructural changes. In this way, the preferred orientation can be estimated by pole figure analysis.

5.4.5 Comparison of the Phase and Crystal Orientation on Whisker-Prone Plating and Whisker-Free Plating

This section presents several examples of the analysis and comparison of phase and crystal orientation on platings that do and do not tend to form mechanically induced whiskers, as shown in Figure 5.35 [40]. All of the maps in Figure 5.35 were obtained before applying mechanical stress. After applying mechanical stress, the following features were observed: nonreflowed Sn–Cu plating is whisker prone, reflowed Sn–Cu plating is whisker and nodule prone, nonreflowed Sn–Ag–Cu plating is nodule prone, and the reflowed Sn–Ag–Cu plating is whisker free.

The left column in Figure 5.35 shows cross-sectional EBSD phase maps of the platings before applying mechanical stress. These maps are overlaid on associated image quality (IQ) maps. IQ maps are grayscale maps that represent the intensity of EBSDs, so they provide useful visualization of microstructural features such as grain boundaries. The colors in the phase maps indicate measured crystalline phases: blue, red, and green represent β-Sn, Ag_3Sn, and $(Cu,Ni)_6Sn_5$ phases, respectively. The crystalline phases of nonreflowed and reflowed Sn–Cu platings were identified as β-Sn.

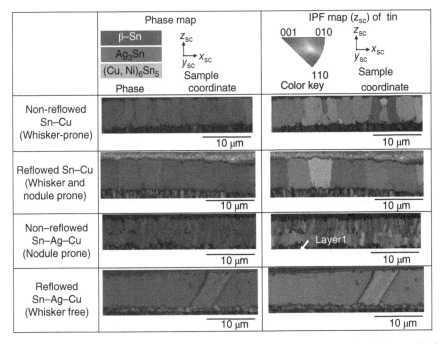

Figure 5.35 Cross-sectional EBSD analysis of the samples before mechanical stress had been applied: left is cross-sectional EBSD phase maps, and right is cross-sectional IPF maps of tin in the sample z-direction (normal to the plating surface; z_{sc}). Source: Reproduced by permission of the Japan Institute of Metals. Please visit www.wiley.com/go/Kato/TinWhiskerRisks to access the color version of this figure.

The nonreflowed Sn–Ag–Cu plating contained Ag_3Sn precipitates that were less than 2 µm in size in the β-Sn phase. The reflowed Sn–Ag–Cu platings contained large (over 5 µm) Ag_3Sn precipitates in the β-Sn phase. Large $(Cu,Ni)_6Sn_5$ precipitates were not detected in the reflowed or nonreflowed Sn–Ag–Cu platings. If they existed, the $(Cu,Ni)_6Sn_5$ precipitates in the platings were probably too small (i.e., smaller than the measurement step size of 0.1 µm) to be detectable by EBSD measurements. The right column in Figure 5.35 shows cross-sectional IPF maps of tin in the sample z-direction (normal to the plating surface; z_{sc}) for the same areas of the phase maps in the left column of Figure 5.35. The colors in the IPF maps indicate the measured Sn grain orientations according to the color key.

These figures clearly show that nonreflowed and reflowed Sn–Cu platings have columnar structures with various grain orientations. The reflowed Sn–Cu plating had larger grains than the nonreflowed Sn–Cu plating. The nonreflowed Sn–Ag–Cu plating had a double-layer columnar structure (layers 1 and 2 in the right column in Fig. 5.35) because each layer was plated independently. Note that the reflowed Sn–Ag–Cu plating had only one layer, with no columnar grains. The tin grains of the reflowed Sn–Ag–Cu plating were much larger than the columnar grains in the other three platings.

This study demonstrated that one method of mitigating mechanically induced Sn whisker formation is to employ the plating with large noncolumnar grains, for which it is difficult to cause microstructural change (including twin formation) by applying mechanical stress. Reflowed Sn–Ag–Cu plating is a very promising whisker-free plating.

In conclusion, this section briefly described several crystal orientation characterization techniques and then focused on the principles of and advances in EBSD for investigating the effects of crystal orientation on tin whisker formation. Several examples of EBSD analysis of the effects of crystal orientations on tin whisker formations were introduced.

To plot a permanent whisker-mitigation strategy, it is insufficient to merely comprehend the whisker formation phenomena. The crystal orientation of a whisker and its surroundings are also very important. It is hoped that significant additional crystal orientation data would permit the formulation of drastically improved whisker-mitigation strategies and should therefore be given a high priority.

5.5 CONCLUSIONS

Many examples of characterization by using TEM, SEM, and EBSD techniques for tin and tin-based alloy finishes in this chapter have helped instruct how to easily clarify the essential characteristics of the films so that we can deepen our understanding of whisker formation mechanisms.

This chapter also showed not only interesting data regarding whisker phenomena and useful characteristics of the films but also fundamental data such as indexed reciprocal lattice planes for typical crystal phases and an EBSD principle and a variety of EBSP measurement data. These should be helpful and provide a good guide to expansion of future whisker studies for the reader.

ACKNOWLEDGMENTS

The information and data regarding TEM, STEM, and 3D analytical holder in Section 5.2 were partially described by courtesy of Hitachi High-Technologies Corporation. We would like to thank Mr Katsutaka Kimura, Mr Tsuyoshi Onishi, Mr Takahito Hashimoto, and Mr Yuichi Madokoro at Hitachi High-Technologies Corporation for their kind support and discussions about descriptions in Section 5.2. We also would like to thank Professor Norihito Sakaguchi and Professor Seiichi Watanabe at Hokkaido University for their collaboration to calculate diffraction patterns described in Section 5.2.

REFERENCES

1. Zworykin VK, Morton GA, Ramberg EG, Hillier J, Vance AW. *Electron Optics and The Electron Microscope*. New York: Wiley; 1948.

2. Cosslett VE. *Introduction to Electron Optics*. Oxford: Clarendon Press; 1950.

3. Klemperer O. *Electron Optics*. Cambridge: Cambridge University Press; 1953.

4. Hall CE. *Introduction to Electron Microscopy*. New York: McGraw-Hill; 1953.

5. Sturrock PA. *Static and Dynamic Electron Optics*. Cambridge: Cambridge University Press; 1955.

6. Ruska E, Wolff O. Ein hochauflosendes 100 kV elektronemikroskop mit kleinfelddurchstrahlung. Z Wiss Mikr 1956;62:465.

7. Leisegang S. Elektronenmikroskope. Handb Phys 1956;33:396.

8. Hirsch PB, Howie A, Nicholson RB, Pashley DW, Whelan MJ. *Electron Microscopy of Thin Crystals*. London: Butterworth; 1965.

9. Kato T, Akahoshi H, Nakamura M, Hashimoto T, Nishimura A. Effects of minor elements in Cu leadframe on whisker initiation from electrodeposited Sn/Cu coating. IEEE Trans Electron Packag Manuf 2007;30(4):258–269.

10. Kato T, Akahoshi H, Nakamura M, Terasaki K, Iwasaki T, Hashimoto T, Nishimura A. Correlation between whisker initiation and compressive stress in electrodeposited SnCu coating on Cu leadframe. CD Proceedings of 4th iNEMI Workshop on Tin Whiskers; May 2007; Reno, Nevada: iNEMI, ECTC, and CPMT; 2007.

11. Kato T, Akahoshi H, Nakamura M, Terasaki K, Iwasaki T, Hashimoto T, Nishimura A. Correlation between whisker initiation and compressive stress in electrodeposited tin–copper coating on copper leadframes. IEEE Trans Electron Packag Manuf 2010;33(3):165–176.

12. Kamino T, Yaguchi T, Konno M, Ohnishi T, Ishitani T. A method for multidirectional TEM observation of a specific site at atomic resolution. J Electron Microsc 2004;53:583–588.

13. Reimer L. *Scanning Electron Microscopy: Physics of Image Formation and Microanalysis* (Springer Series in Optical Sciences). Springer; 1998.

14. Wells OG. *Scanning Electron Microscopy*. McGraw-Hill Inc.; 1975.

15. Mizuguchi Y, Murakami Y, Tomiya S, Asai T, Kiga T, Suganuma K. Effect of crystal orientation on Kink formation on mechanically induced Sn whiskers. J IEICE (C) 2012;J95-C(11):333–342, [in Japanese].

16. Suganuma K, Baated A, Kim K-S, Hamasaki K, Nemoto N, Nakagawa T, Yamada T. Sn whisker growth during thermal cycling. Acta Mater 2011;59(19):7255–7267.

17. Osterman MD. Tin whisker growth measurements and observations. Proceedings of 2nd International Symposium on Tin Whiskers, 1-1-1, 2008.

18. JEDEC Standard, JESD22-A121. *Test Method for Measuring Whisker Growth on Tin and Tin Alloy Surface Finishes.* JEDEC; 2008.

19. JEDEC Standard, JESD201A. *Environmental Acceptance Requirements for Tin Whisker Susceptibility of Tin and Tin Alloy Surface Finishes.* JEDEC; 2008.

20. IEC Standard, 60068-2-82, Environmental testing – Part 2–82: Tests – Test XW1: Whisker Test Methods for Electronic and Electric Components, Edition 1.0, 2007.

21. Lee B-Z, Lee DN. Spontaneous growth mechanism of tin whiskers. Acta Mater 1998;46(10):3701–3714.

22. Furuta N, Hamamura K. Growth mechanism of proper tin-whisker. Jpn J Appl Phys 1969;8(12):1404–1410.

23. Baker GS. Angular bends in whiskers. Acta Mater 1957;5:353–357.

24. Yang F, Li JCM. Deformation behavior of tin and some tin alloys. J Mater Sci Mater Electron 2007;18:191–210.

25. Bieler TR, Jiang H, Lehman LP, Kirkpatrick T, Cotts EJ, Nandagopal B. Influence of Sn grain size and orientation on the thermomechanical response and reliability of Pb-free solder Joints. IEEE Trans Compon Packag Technol 2008;31(2):370–381.

26. Honda K. Configurations, formation and burgers vector of coupled dislocations straddling a polygon wall in white tin single crystals. Jpn J Appl Phys 1969;8(12):1404–1410.

27. Wagner RS, Ellis WC. Vapor–liquid–solid mechanism of single crystal growth. Appl Phys Lett 1964;4(5):89.

28. Wagner RS, Ellis WC. The vapor–liquid–solid mechanism of crystal growth and its application to silicon. Trans Metall Soc AIME 1965;233:1053–1064.

29. Mizuguchi Y, Murakami Y, Tomiya S, Asai T, Kiga T. Whisker and nodule formation on lead-free tin plating by external stress. Proceedings of 2nd International Symposium on Tin Whiskers, 2-2-6, 2008.

30. Mizuguchi M, Murakami Y, Tomiya S, Asai T, Kiga T, Suganuma K. Effect of crystal orientation on mechanically induced Sn whiskers on Sn–Cu plating. J Electron Mater 2012;41(7):1859–1867.

31. Shibutani T, Yu Q, Shiratori M, Pecht MG. Pressure-induced tin whisker formation. Microelectron Reliab 2008;48:1033–1039.

32. Osenbach JW, Shook RL, Vaccaro BT, Potteiger BD, Amin AN, Hooghan KN, Suratkar P, Ruengsinsub P. Sn whiskers: material, design, processing, and post-plate reflow effects and development of an overall phenomenological theory. IEEE Trans Electron Packag Manuf 2005;28(1):36–62.

33. Fisher RM, Darken LS, Carroll KG. Accelerated growth of tin whiskers. Acta Mater 1954;2:369–373.

34. Yang F, Li JCM. Deformation behavior of tin and some tin alloys. J Mater Sci Mater Electron 2007;18:191–210. Reimer, Scanning Electron Microscopy: Physics of Image Formation and Microanalysis (Springer Series in Optical Sciences) (Springer; 2nd completely rev. and updated ed., 1998).

35. Alam MN, Blackman M, Pashley DW. High-angle Kikuchi patterns. Proc R Soc A 1954;221(1145):224–242.

36. Weilie Z, Zhong Lin W, editors. *Scanning Microscopy for Nanotechnology*. New York: Springer; 2010.

37. Venables JA, Harland CJ. Electron back-scattering patterns – a new technique for obtaining crystallographic information in the scanning electron microscope. Philos Mag 1973;27(5):1193–1200.

38. Terang AU, Bieler TR. The orientation imaging microscopy of lead-free Sn–Ag solder joints. JOM 2005;57(6):44–49.

39. Maruyama S, Kiho H. Intersections of {301}, {101} twin bands in tin. J Phys Soc Jpn 1956;131(5):516–521.

40. Mizuguchi Y, Murakami Y, Tomiya S, Asai T, Kiga T, Suganuma K. Effect of crystal orientation on Sn whisker-free Sn–Ag–Cu plating. Mater Trans 2012;53(12):2078–2084.

6

OVERVIEW OF WHISKER-MITIGATION STRATEGIES FOR HIGH-RELIABILITY ELECTRONIC SYSTEMS

DAVID PINSKY

Raytheon Integrated Defense Systems, Tewksbury, Massachusetts, USA

6.1 OVERVIEW OF TIN WHISKER RISK MANAGEMENT

In practice, management of the risks associated with tin whiskering is not fundamentally different from managing risks associated with other failure modes that bedevil electronic systems. However, the underlying uncertainties and associated lack of predictability of the tin whiskering phenomenon create a challenging environment for risk management. Appropriate management of these risks requires each of the various stakeholders to address those particular challenges that are associated with their roles.

A partial listing of the risk management activities associated with various critical stakeholders is summarized in Table 6.1. An overview of these activities is provided in the following sections. Technical details of the specific approaches outlined in the section are provided in the body of this chapter.

6.1.1 Metal Finisher/Component Packager

Typically, the composition of the metal finish will be provided as a requirement to the metal finisher. Clearly, if these finishes are not tin-based, there will be no concern

Mitigating Tin Whisker Risks: Theory and Practice, First Edition.
Edited by Takahiko Kato, Carol A. Handwerker, and Jasbir Bath.
© 2016 John Wiley & Sons, Inc. Published 2016 by John Wiley & Sons, Inc.
Companion website: www.wiley.com/go/Kato/TinWhiskerRisks

TABLE 6.1 Partial Listing of the Risk Management Activities Associated with Various Critical Stakeholders

Stakeholder	Risk Management Activities
Metal finisher/component packagers	Selection of lower risk tin-plating solutions
	Proper process control of plating operations
	Performance of whisker propensity tests
Component packaging designer	Selection of lower risk finishes
	Specification of whisker propensity testing
	Geometric design to reduce whisker risk
Assembly designer	Selection of components with lower risk finishes
	Specification of coatings and barriers
	Specification of component separations
Subsystem assembler	Verification of component finishes
	Control of solder coverage
	Control of coating thickness and coverage
Systems designer and integrator	Identification, flow down, and verification of appropriate tin whisker risk management requirements for subsystem providers
	Identify root cause of failures occurring during system integration or reported by end users
	Incorporation of redundancy and fault tolerance into system design
End user	Collect failure event and reliability data

about tin whisker risk. If the specified finish is tin-based, then the metal finisher should take actions to minimize the risk of subsequent whisker growth.

The physical properties of the deposit that affect whiskering propensity are discussed in detail in other chapters of this book. In practice, the metal finisher will need to select from among commercially available plating baths and associated recommended process conditions. The detailed chemistry of commercial plating baths has been subject to rapid evolution, which is a trend that is likely to continue. These chemistries and processes are being driven to optimize cost of the solutions, speed of processing, solderability of the deposit, and tin whiskering propensity. As is usual with such optimizations, trade-offs are involved. Therefore, metal finishers placing a high priority on reduced whiskering propensity will likely need to make compromises in some of these other important areas.

In some cases, the direct customer of the metal finisher will require that whisker propensity testing be performed, typically in accordance with specification JEDEC JESD 211 [1].

Since this testing requires 6 months to execute, it is more often the case that this testing is performed by the manufacturer of the plating bath systems, as a form of qualification of the resistance to whiskering of their particular products. There are currently systems on the market that are advertised to have passed JESD 211 testing [1] with no whiskers reported. (Growths shorter than 10 μm are not considered to be whiskers for this purpose.) Use of such a plating system therefore provides the *possibility* of creating deposits that exhibit very low propensity for whisker growth. However, differences between the process parameters used for the actual product and those used for the original qualification of the bath by its supplier can result in disparate whiskering propensities. For example, impurities introduced into the system during component processing can have a major effect on the whiskering propensity of the resulting deposit. Therefore, careful control of the process parameters and chemistries are required if the advertised whiskering propensity is to be reproduced in actual product.

6.1.2 Component Packaging Designer

The component packaging designer typically has control over specification of the finishes to be used. Indeed, if all component packaging designers could be convinced to specify only non-tin-based finishes, the entire issue of tin whiskers could be made to vanish. Nonpure tin, RoHS-compliant finish options are available, including NiPdAu, NiAu. However, these finishes tend to be more expensive than pure tin and are not practical for every package geometry. As a result, tin-based finishes are very commonly specified.

Generally speaking, a line of sight must exist between terminations for there to be a meaningful risk of tin whisker shorting. In addition, the risk of whisker shorting decreases significantly with increasing gap dimensions. Also, those portions of device terminations that are wetted by solder exhibit significantly reduced propensity for whiskering. Therefore, bottom-only terminations and other very-low-profile terminations will be provided with this mitigation during next higher assembly.

A recent trend in small footprint packages involves exposed metallization on the bottom, which is expressly forbidden to be soldered or otherwise contacted electrically to other terminations. This creates a special, difficult-to-mitigate tin whisker risk. It is therefore clear that package geometry has a significant effect on the possibility of whisker-induced failures. Unfortunately, considerations other than whisker-induced unreliability typically drive package geometries (thermal management, signal integrity, footprint minimization, etc.). As a result, considerations of these geometric factors are typically of more utility to the subsystem designer during the selection of components than to the component designer.

6.1.3 Assembly Designer

The designer of the assembly (often a circuit card assembly but not always) will need to manage numerous requirements, of which tin whisker risk control will be but one. The optimum choice for tin whisker risk mitigation will be to select components that do not utilize pure tin finishes. Some components are available in standard finishes such as nickel–palladium–gold finishes, which are both RoHS-compliant and non-tin-based. For applications where compliance to RoHS is not a requirement, components with tin–lead finishes can also be selected when available. Unfortunately, pure tin will likely be the only practical choice for the finishes of numerous components on any given assembly, so additional mitigations will need to be incorporated. Since bulk solder is far less susceptible to whisker growth than are pure-tin-plated terminations, the selection of ball grid array packages in place of components with leads is another viable strategy. Terminations that are completely covered by solder during SMT processing will also provide for lower overall tin whisker risk. Therefore, components with bottom-only terminations are also preferred. During SMT processing, solder will wick up to coat some portion of terminations at a distance above the seating plane of the device. The amount of this wicking will be a function of the solder process details and the geometry of the pad and the component termination. A subset of components will be sufficiently small for a given process that their terminations will be fully covered by solder. Parts of this type have been described as "self-mitigating." The relationship between component geometry, solder process, and solder coverage for this "self-mitigation" of pure-tin-finished parts by tin–lead solder has been described. Selection of components of this type is also preferred.

The gap distance across which the tin whisker must bridge is a critical risk factor. Therefore, increasing the spacing between tin-plated surfaces and adjacent conductors is a very effective means of reducing tin whisker risk. Unfortunately, small gap distances often occur between adjacent terminations of a given component, which is impossible for the assembly designer to affect.

Where it is necessary to use pure tin finishes in close proximity, without the benefit of solder coverage, the use of physical barriers can be used to reduce the tin whisker risk. The most common types of barriers that are applied are conformal coating for circuit card assemblies and solid insulation for cable and other interconnection assemblies. Potting can also be a very effective mitigator, but it involves many drawbacks, so it is rarely employed solely for the purpose of tin whisker risk mitigation. However, in those instances where the use of potting is desired for other purposes (thermal management, high-voltage insulation, etc.), it can also be relied upon to provide excellent tin whisker risk mitigation.

Designers must remember that all components within an electrical system can induce tin whisker risk, not only the electrical components. Nonelectrical components such as heat sinks, card rails, EMI shields, etc., have been known to grow whiskers and induce failures. In fact, these mechanical components tend to have much larger tin-plated surfaces than electrical components and can be much larger in dimension. Therefore, they often create much higher overall risk of tin whisker unreliability than are posed by the electrical components. Typically, designers have greater latitude in selection of the finishes of such components than with purchased standard electronic parts, so there are correspondingly greater opportunities for designers to reduce tin whisker risk in this area through the selection of alternative finishes.

6.1.4 Subsystem Assembler

The subsystem assembler, typically a circuit card assembly facility, will need to reliably implement the designs of the assembly designer so that the intended tin whisker risk mitigations are achieved. It is assumed that any competent subsystem assembler will be capable of manufacturing assemblies that are compliant with the prevailing industry requirements such as IPC-A-610. However, these requirements typically do not encompass the controls that are generally preferred for achieving optimum tin whisker risk mitigation. Therefore, additional controls may be required.

When coverage of tin-plated terminations is intended for the purpose of tin whisker risk mitigation, it will be necessary to validate that this coverage is effectively achieved by the given combination of process, component geometry, pad geometry, and stencil geometry. The formation of solder joints compliant to the applicable class of IPC-A-610 may not be sufficient to assure that this desired level of coverage is achieved.

Conformal coating has long been used to suppress corrosion-induced short circuits on circuit card assemblies. The principal mechanism involved is voltage-driven metal migration on the surface of the card between parallel conductors. Validation of conformal coat performance has therefore typically been achieved through surface insulation resistance testing, and the thickness requirements for coating deposition have typically been specified for the flat surface of the cards. In contrast, tin whisker risk mitigation is principally provided by coating the surfaces of leads including their backs and corners.

Coating processes that provide for ample coverage on flat horizontal surfaces may not necessarily provide coverage on corners and shadowed regions of vertical surfaces. Therefore, coating material and process combinations that conform to all of the traditional performance requirements may not necessarily provide optimum tin whisker risk mitigation. The uses of vapor-deposited coatings such as parylene provide an advantage for tin whisker risk mitigation because of their ability to uniformly cover all surfaces. Unfortunately, such coatings are not always practical for all applications. Whatever coating process and material combinations are selected, care must be taken to assure that adequate coverage is achieved in the required locations.

6.1.5 System Designer/Integrator

The top-level system designer/integrator will have responsibility to assure the overall performance including reliability and global responsibility for management of the impacts of tin-whisker-induced unreliability. Management of these risks will require an assessment of the potential impacts of tin-whisker-induced unreliability with respect to system level: mission requirements, life requirements, maintenance requirements, etc.

This process will occur in two general steps: assessing the criticality of each subsystem based upon top-level requirements, then imposing a commensurate tin whisker risk mitigation requirement to each subsystem. In situations where acceptable levels of unreliability may not be achievable for individual subsystems, the use of redundancy and other forms of fault tolerance may be required.

System integrators should make adequate provisions for the detection and identification of tin-whisker-induced failures that may occur during system integration. Since tin-whisker-induced unreliability can be difficult to identify, it is recommended that test and failure analysis personnel be provided with tin whisker awareness training.

6.1.6   End User

It is the end user who is most likely to suffer the consequences of any tin-whisker-induced unreliability. Frequently, tin whisker effects result in transient failures that can be difficult to isolate. Technicians and engineers who perform troubleshooting and repair activities at or with end users should also receive tin whisker awareness training.

6.2   DETAILS OF TIN WHISKER MITIGATION

For electronic equipment, tin whiskers present reliability risk due to their ability to create electrical shorts, initiate plasma arcs, and create mechanical obstructions. While the exact process of whisker formation has not been fully explained, it is clear that various factors can produce whisker growth on tin-finished surfaces.

In order to reduce failure risk presented by tin whiskers, electronic equipment manufacturers need to understand various mitigation strategies and their effectiveness at preventing whisker-related failures. Mitigations can be broadly separated into part selection/specification options and equipment manufacturer mitigations. Part selection/specification options relate to the ability of the equipment manufacture to obtain parts with specific mitigation strategies. This may include selecting a specific lead-free tin surface finish, selecting a specific plating thickness, selecting finishes with an underlayer coating, selecting plating that have been qualified to a particular standard, or selecting parts that have been subject to specific treatment conducted by the part supplier that is expected to mitigate whisker growth.

With regard to equipment manufacture mitigation strategies, these strategies include operations that can be conducted or contracted by the equipment manufacturer on parts after they have been procured for the supply chain. Equipment manufacture mitigation strategies may include setting spacing limits, hot solder dipping, solder assembly and inspection, encapsulation, or application of an insulation coating layer.

With regard to whisker growth, research clearly indicates that all tin finishes are capable of producing whisker growth, including tin–lead. For past electronics, the level of whisker formation that occurred on tin–lead finished surfaces was found to be acceptable. The difficulty in for any new part-level mitigation strategy is to prove that it provides the same level of mitigation as tin–lead. This difficulty arises from the inability to develop tests and inspection mechanisms that can reliably predict whisker growth that may occur in the field. Each of the part selection/specification options will be discussed.

Alloying lead with tin has been shown to significantly reduce whisker growth propensity. The effectiveness of alloying Pb with Sn to reduce whisker formation was studied by Arnold from Bell Laboratories [2]. In this study, Arnold reported on the beneficial effects from alloying tin (Sn) films with lead (Pb) in amounts ranging from 3% to 10% by weight. With the ban on use of Pb in electronics, research has been conducted on other tin alloys. The actual mitigation provided by the addition of lead remains unclear. Some have attributed the mitigation due to the formation of equiaxial grains in a tin–lead plating process. Others have attributed the lead phase as reducing the stress buildup in the film. It should be noted that tin whiskers rarely have been documented on tin–lead surfaces [3].

Tin copper has been used by a number of part manufacturers. Many of these manufacturers have test data that indicates limited whisker growth. However, NIST reported on the influence for copper contamination on tin films and found the addition of trace amounts of copper-induced large whisker growth [4]. Bismuth has also been proposed as an alloy component with tin to reduce whisker formation.

6.2.1 Substrate Selection

For electronic components, copper alloy and nickel–iron substrates are often used for lead-frame material. The substrate on which the tin finish is applied has been demonstrated to influence whisker formation. Copper substrates form copper–tin intermetallics with tin. The formation of copper–tin intermetallics has been attributed to increasing stress in the plating due to an increased volume of the copper–tin intermetallic. For brass substrates, which contain copper and zinc, it is speculated that zinc, similarly to copper, increases the stress levels in the plated tin and promotes whisker formation [5–7]. Brass is the substrate of choice for researchers seeking to create whisker growth specimens.

Researchers have found similar growth rates for tin plated onto zinc [5]. Nickel–iron (Alloy 42) substrate does not readily form extensive intermetallics, and the intermetallics have been lower in volume than the constituents, which result in lowering or creating a tensile stress state in the tin-plated film. While Alloy 42 substrates do not readily form whiskers under ambient conditions, the coefficient of temperature expansion (CTE) of Alloy 42 is approximately 5 ppm/°C. For tin, which has a CTE of approximately 23 ppm/°C, a large mismatch of temperature expansion exists with Alloy 42, unlike copper, which has a CTE of 17 ppm/°C. Under temperature-cycle conditions, tin on Alloy 42 can form whiskers. Endo et al. [7] found that tin deposited on copper, nickel, and silver substrates did not produce whiskers in over 2 years of storage observation.

6.2.2 Nickel Underlayer

To remove the whisker growth path due to copper–tin intermetallics, an intermediate layer of metal can be plated between the base copper substrate and tin. Nickel has been one of the most popular materials to be used as a barrier layer. The reduction of tin whisker formation on nickel substrates has been documented by multiple researchers [7–9]. It has been suggested that nickel prevents copper from entering the tin plating and that the nickel–tin intermetallic compound formed reduces the stress state in the tin layer. Hada et al. [9] found no whisker growth on 8 μm tin deposit over a 2 μm Ni deposit on copper. Endo et al. [7] found that a nickel underlayer of 0.5–1.5 μm over brass was effective in eliminating whisker growth on stored samples.

6.2.3 Plating Thickness

The thickness of the tin deposit has been demonstrated to effect whisker growth [5, 9]. Glazunova [5] found whisker growth on copper to be a maximum on tin deposits with thicknesses between 2 and 5 μm and whisker growth on brass was a maximum on tin deposits with up to 20 μm thickness. Hada et al. [9] showed that 2 μm deposits of tin on copper had much longer whiskers than those found on 10-μm-thick tin deposits.

6.2.4 Annealing

Annealing is a heat treatment wherein a material is altered, relieving residual stress and changing grain structure. Heat treatment of tin has been used to mitigate tin whisker growth. Multiple electronic part manufacturers report using an annealing step of 150°C for 1 h within 24 h of plating as a whisker-mitigation process [10–12].

Lee and Lee [13] demonstrated that a 150°C 1 h annealing step resulted in an initial tensile stress in a plated tin layer, which reduced to near-zero stress with no observable whisker growth. A similar tin deposit that did not receive the postbake treatment had an initial tensile stress but developed a compressive film stress of approximately 8 MPa, which reduced within 50 days to approximately 5 MPa with visible whisker growth. Hada et al. [9] showed that heat treatments of 140°C for 30 min had 3 h delayed whisker formation of tin deposits on copper substrates, but the treatment did not eliminate whisker growth.

In examining a 2-μm-thick tin deposit and a 10-μm-thick tin deposit, the 3 h annealing step delayed whisker growth for over a year but whiskers longer than 1.5 mm were observed on the 10 μm deposit after 500 days and whiskers longer than 2 mm were observed on the 2 μm deposit in less than 400 days. Fukuda et al. [14] findings in testing tin over copper annealed at 150°C for 1 h support the finding that annealing may delay but does not eliminate whisker growth. Test results from JEITA whisker studies with over 20,000 h of observation also confirm that annealing delays but does not eliminate whisker growth [15].

6.2.5 Solder Dipping

Solder dipping of tin-finished terminations with tin–lead solder can be used to convert the plated tin finish into a hot-dipped tin–lead finish. Since tin–lead finishes exhibit negligible propensity for whisker growth, this is a very effective mitigation. The principal concern with this approach is that the molten solder must cover the entire length of the terminations, right up to the component body. This can induce stresses in the component during solder dip that are in excess of those encountered during normal solder attach, raising a concern that components may be damaged. Robotically controlled hot-solder dip processes have been developed that minimize the potential for such damage. An industry specification, SAE GEIA-STD-0006, has been published that can be used to control such processes.

6.2.6 Solder Coating during SMT

The portion of terminations that are wetted by solder during normal SMT attach processes will no longer exhibit the as-plated finish. If the resulting finish is eutectic tin–lead solder, then the risk of tin whiskers is fully mitigated in that region. If the resulting finish is a lead-free solder, then the risk of tin whiskers may be reduced but not eliminated [16].

The portion of the terminations that are not coated with solder will have no benefit from this mitigation. In some instances, the solder will entirely replace the original tin plating, fully mitigating the whisker risk. This type of coverage is not a requirement of IPC-A-610 standard, so it cannot be inferred to occur in general. Conditions under which components will be fully mitigated have been investigated, and the ability to mitigate was found to depend upon the geometry of the terminations [17].

6.2.7 Coating

Conformal coating, which is used to provide protection of exposed metal surfaces and electronic devices on printed wiring boards, has also been considered to mitigate tin whisker failures. Conformal coating provides two types of whisker failure mitigations, which are depicted in Figure 6.1. In the first case, conformal coat can prevent an incoming whisker from contacting an exposed metal surface. In the second case, a conformal coat can contain whiskers, preventing whiskers from escaping the surface and bridging to an adjacent surface. In both cases, coverage of the surfaces is an important consideration. For some spray applications, high viscosity of the coating and surface tension of the film can cause very thin coverage at the edges of surfaces,

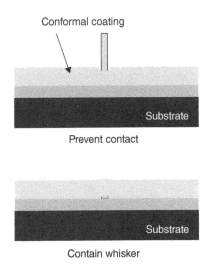

Figure 6.1   Conformal coating tin-whisker-mitigation modes.

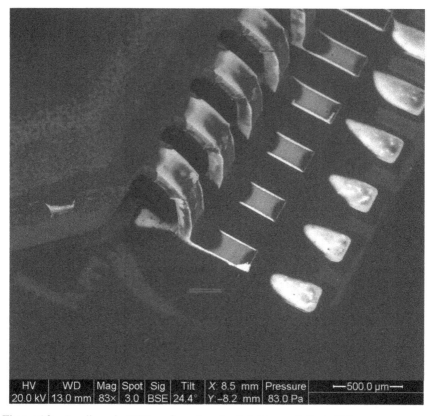

Figure 6.2   Acrylic resin (AR) conformal coating bright areas indicate little or no coverage.
Source: Image courtesy of Center for Advanced Life-Cycle Engineering [CALCE].

with thicker coverage near the center of the surface. Figure 6.2 depicts a coverage
issue that can occur with liquid-sprayed conformal coats.

When a surface is completely covered with a conformal coat, the ability of a tin
whisker to contact the covered surface is dramatically decreased. It has been estimated
that most whiskers arriving from a distance will bend and buckle rather than penetrate
the coating and create a short [18].

6.2.7.1 *Conformal Coat Model* A Model to Predict the Mitigation of the Effects
of Conformal Coating

The model described here has been constructed based upon three assumptions:

- Conformal coat on real circuits exhibits a certain amount of voids,
- Conformally coated tin-plated surfaces will grow whiskers at a reduced incidence,

- Whiskers cannot penetrate a conformally coated surface at a distance from the origin of the whisker.

Consider two identical parallel plates, each covered with a layer of tin and a layer of conformal coat that exhibits incomplete coverage. Let v represent the fraction of each surface that is void of conformal coat.

Let $P(u)$ represent the probability that a whisker will form per unit area of the surface that is uncoated and grow sufficiently that it eventually comes into contact with the opposing surface.

Let $P(c)$ represent the probability that a whisker will form beneath a unit area of the surface that is coated, penetrate the coating, and grow sufficiently that it eventually comes into contact with the opposing surface.

The probability that a whisker will grow from one of the plates(1), bridge the intervening space, and contact the opposing plate(2) is given by

$$P(1 \rightarrow 2) = (1 - v)P(c) + vP(u) \qquad (6.1)$$

The probability is identical for whiskers bridging in the opposite direction, by symmetry.

It is assumed that whiskers that bridge the gap between the plates and eventually contact a portion of the opposing plate that is covered by conformal coating will not result in electrical contact. The model is illustrated in Figure 6.3.

There are two distinct modes by which electrical shorting could occur: (1) a whisker bridges the gap and contacts a region of the opposing surface that is not coated; or (2) whiskers grow simultaneously from opposite sides and contact each other in the space between the plates.

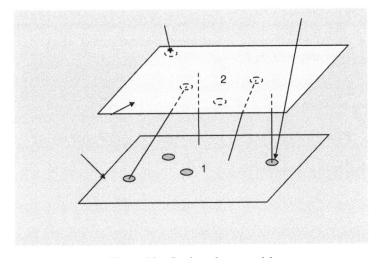

Figure 6.3 Conformal coat model.

Let P_1 represent the probability of a whisker contacting a portion of the opposing surface that is void of conformal coat. This is given by

$$P_1 = v[(1 - v)P(c) + vP(u)] \qquad (6.2)$$

Let P_2 represent the probability that two whiskers grow from opposite surfaces, to meet in the middle. Precise estimation of this probability will be very difficult. It will depend upon the probability that a pair of whiskers will grow from opposite sides, exhibiting complementary lengths and orientations. Monte Carlo techniques could be employed to make an estimation, but this will be complicated by the fact that whiskers move over time and that they might be attracted to one another through electrostatic force. For estimation purposes, we will approximate that the probability that such a pair of whiskers will grow is proportional to the probability that a pair of bridging whiskers will grow. The probability is estimated by

$$P_2 \approx F[(1 - v)P(c) + vP(u)]^2 \qquad (6.3)$$

where F is a term of unknown form that accounts for the particulars of the geometry and any mutual affinity that the whiskers may exhibit. It is assumed that this term does not depend upon either $P(c)$ or $P(u)$.

Let P_T represent the total probability that a short will form due to whisker growth on either surface, either contacting a void or another whisker. This is given by the sum

$$P_T = 2(P_1 + P_2) \qquad (6.4)$$

By combining Equations 6.2–6.4,

$$P_T = 2\{v[(1 - v)P(c) + vP(u)] + F[(1 - v)P(c) + vP(u)]^2\} \qquad (6.5)$$

Let E represent a quantity called the "coating effectiveness" This quantity is the factor by which the conformal coat suppresses the formation of whiskers exhibiting bridging geometries. That is,

$$E = P(c)/P(u) \qquad (6.6)$$

Replacing $P(c)$ with the value $EP(u)$, Equation 6.5 becomes

$$P_T = 2\{[v(1 - v)]EP(u) + v^2P(u)\} + F\{(1 - v)EP(u) + vP(u)\}^2 \qquad (6.7)$$

By algebra from Equation 6.6,

$$P_T = 2P(u)(Ev - Ev^2 + v^2) + 2FP^2(u)(E - Ev + v)^2 \qquad (6.8)$$

In the case of a void-free coating, the value of v goes to zero, and the equation reduces to

$$P_T = 2FE^2P^2(u) \qquad (6.9)$$

In the case of a coating that provides perfect whisker suppression, but exhibits voids, the value of E goes to zero and the equation reduces to

$$P_T = 2v^2 \left[P(u) + FP^2(u) \right] \qquad (6.10)$$

In a case where the coating provides substantial yet imperfect whisker suppression and exhibits a nonzero, but small void fraction, the following approximation holds:

$$v \ll 1 \text{ and } E \ll 1$$

In this case, the higher order terms become insignificant, and Equation 6.8 can be approximated by

$$P_T \approx 2P(u)(Ev + v^2) + 2FP^2(u)(E^2 + v^2) \qquad (6.11)$$

If it is also the case that the probability that a whisker will form from an uncoated surface that is sufficient to bridge the gap is also very small, the implication is that

$$P(u) \ll 1$$

If this assumption is also true, then Equation 6.11 can be reduced even further to

$$P_T \approx 2P(u)(Ev + v^2) \qquad (6.12)$$

The factor in Equation 6.12, $Ev + v^2$, provides an approximation of the overall reduction in the probability of creating a short that can be gained by an application of conformal coat that exhibits a void fraction of v and a "Coating Effectiveness" of E (providing that the small-value assumptions listed earlier are valid). This new factor depends upon both the physical properties of the conformal coat, which determine E, and how the coating is applied, which determines the void fraction v. The portion of the term that accounts for whiskers growing through the conformal coat is given by Ev, while v^2 accounts for whiskers that grow from an uncoated surface.

6.2.7.1.1 Model Predictions Consider the case of conformal coat that can suppress whisker growth by a factor of 10:1 and is applied with a surface area void fraction of 5%. Inserting these values into Equation 6.12, an overall reduction in the probability of whisker shorts is predicted to be only 0.0075 of the probability without the conformal coat. From this example, it can be seen that conformal coats that exhibit fairly mediocre properties can still provide significant amounts of risk mitigation.

A higher quality coating that provides 100:1 effectiveness, while exhibiting only a 0.5% void fraction, is predicted to reduce the incidence of whisker shorts to only 0.0000525 of the original value. These are not unreasonable goals for a practical coating.

6.2.7.1.2 Discussion of Model Assumptions One critical assumption is that the overall whisker density is such that the probability of whisker pairs touching in the middle can be ignored. Whisker density measurements have been reported over a very wide range. Specimens that have been prepared especially to produce large quantities of long whiskers have been reported to grow whiskers at densities exceeding $2000\,mm^{-2}$. Measurements performed on platings more typically encountered in production have yielded results in the range of between 50 and $150\,mm^{-2}$ [19].

A typical SOIC lead exhibits a surface area along its edge of approximately one half of a square millimeter. Using the results discussed earlier for the whisker-prone plating, there could be many tens of whiskers growing from the side of one such lead with a length that exceeds 50% of the gap size. Clearly, whisker-to-whisker shorting should not be ignored, for such a plating. If one considers the "whisker-mitigated" plating described by Hilty and Corman [20] or the plating analyzed by Fang, Osterman, and Pecht [21], the expected value for a whisker with a length greater than 50% of the gap size to grow on the side of such a lead will be much less than one. For platings of this sort, the risk of whisker-to-whisker shorts will be negligible.

It is assumed that the probabilities for whisker formation, coating voids, and coating destinations are all uniformly distributed and independent of one another. This is a reasonable assumption for infinite, uniform, parallel plates. For nonidealized geometries, the model will not be too bad so long as the probabilities remain independent. If, however, the probabilities are strongly dependent on one another, the model will break down. That is to say if the same location is at an elevated risk for both a coating void and whisker growth, or for both being a whisker destination and exhibiting a coating void, the incidence of failure will be higher than that the model would predict.

One critical example of this is the sharp edge of a formed lead, which is more likely to protrude through the coating and may also be more likely to form whiskers. It is therefore recommended that the ability of the conformal coat to cover the corners be part of any evaluation performed on coatings intended for whisker mitigation.

The circuit application of most common concern is that of adjacent tin-plated leads on a single electronic package. In this case, the parallel surfaces will not be large in extent. The difference between finite leads and large parallel surfaces is that some whiskers that would have contacted with a large neighbor will miss a small one. Since the model only compares the difference between coated and uncoated performance, this approximation should have little or no effect on the general form of the results. In other words, for both the coated and uncoated cases, we are considering only those whiskers that actually contact the neighbor.

The application of this model to actual circuit configurations requires that data be collected on both the actual void fraction of the conformal coat and the ability of the conformal coat to suppress whisker growth given its actual thickness range. The void fraction will depend upon coating material and process and also upon the details of the electronic package whose leads are being coated. Assessments of this type are currently being planned. It is recommended that investigators collect data of this type so that better predictions can be made.

The concern about whiskers that "tunnel" beneath a conformal coat needs to be considered in light of the actual geometry being evaluated. For the case of adjacent leads of a single microelectronics package, the region of concern for this phenomenon is limited to the portion of the lead that is immediately adjacent to the package. This region will represent a small fraction of the total surface area of the leads, which would tend to reduce the probability that this will be a problem.

If the geometry of interest is the adjacent tin-plated, which runs on the surface of a coated circuit card, the opportunity for whisker "tunneling" will be significantly larger than in the case of adjacent leads, where the coating does not bridge the gap.

6.3 MANAGING TIN WHISKER RISKS AT THE SYSTEM LEVEL

Tin-whisker-induced unreliability is just one of many sources of unreliability that the systems designer must manage. This risk is introduced at the component level, so it shares many aspects of other component-level risks. The basic decision that the systems designer needs to make is the following: how much emphasis and effort must be applied against tin whisker risk mitigation? A simple rule-of-thumb answer to this question is as follows: the effort should be consistent with the effort expended on other component-reliability issues. Following this basic tenet, it follows that the space systems designer that is imposing lot-by-lot qualification and serialization of all components including passive devices should be expending significant levels of effort on tin whisker risk mitigation. Conversely, a system designer that is relying heavily on commercial off-the-shelf (COTS) assemblies with no control over their internal components should be expending minimal levels of effort on tin whisker risk mitigation.

In many instances, specific controls will be mandated by a customer, through statements of work or other vehicles. Subcontractors should carefully review any tin whisker risk mitigation requirements and check to see if they are in keeping with the aforementioned rule of thumb. If they are not, then it is certain that either the tin whisker requirements or the other component requirements are inappropriate, and discussions should be initiated with the customer immediately so that the program can avoid unnecessary costs and reliability risks.

This basic fact that the appropriate amount of effort expended on tin whisker risk management will vary dramatically between systems has been embodied in the industry standard SAE GEIA-STD-0005-2, which defines five distinct control levels for use in managing tin whisker risks. These five levels were created to facilitate standardized requirements management between customers and subcontractors. The standard requires that the design agent produce a tin whisker risk mitigation plan defining the specific controls that will be used to manage tin whisker risk within the framework of the five levels. It is not unusual for multiple control levels to be applied for larger more complex systems. A complete enumeration of the requirements should be obtained by reference to the latest version of the standard. A brief summary of the five control levels is provided in Table 6.2.

Each designer/manufacturer must align their internal processes and procedures to accommodate requirements of the control level or levels that will be invoked. These processes will fall into three broad categories:

TABLE 6.2   GEIA-STD-0005-2 Control-Level Descriptions

Control level	Description	Comments
1	No controls	Not generally used in high-reliability systems
2A	The use of tin is not restricted; some analysis is required to justify lack of restrictions.	Typically employed in high-reliability systems that incorporate significant levels of redundancy or in nonmission critical applications such as test equipment and prototypes.
2B	The use of tin must be justified in each application. However, broad classes of tin applications can be defined by design rules and tin can be used in accordance with them without special permission or notifications.	Used in a wide variety of high-reliability applications.
2C	The use of tin must be justified in each application, which must be individually reviewed and approved	Used in very critical high-reliability applications
3	The use of tin is not permitted under any circumstances.	Very limited use, mostly space-based systems.

- Developing justifications for Level 2A controls
- Defining and implementing methodologies and rules to assess the adequacy of mitigation for individual tin applications for Levels 2B and 2C
- Defining and implementing control methods to assure that components used in applications where tin will not be permitted are in fact not received with tin plating.

6.3.1   Developing Justifications for Level 2A

Under Level 2A controls, the designer will expend no special efforts to assure that the tin is not introduced into the system, nor will attempt to assess the risks posed on an application by application basis. Justifications for pursuing this path will need to involve considerations at higher than the component level.

One common application for Level 2A is for major COTS subassemblies such as routers, displays, and power supplies. One viable approach in such an instance is to obtain meaningful reliability data from the supplier and then verify that this demonstrated performance meets the system-level requirements imposed on the subassembly in your application. The logic is straightforward – if the overall unreliability is acceptable, then the portion of the unreliability due to tin whiskers must also be acceptable. The challenge here is to obtain the type of data that is required from the supplier. Newly introduced models may not have sufficient field history behind them to generate meaningful mean time between failures (MTBF) data. Operating conditions during which supplier MTBF data was obtained may not be a good match for the intended operating conditions by the specific end user. These challenges are really just a subset of the challenges that are faced by high-reliability system integrators who rely on COTS subsystems.

Another common use for this control level is for noncritical hardware such as test systems, prototypes, and engineering models. The basic rationale for these systems is that intermittent failures, although inconvenient, are unlikely to result in catastrophic consequences. Eliminating tin controls under such circumstances can provide much appreciated relief for engineers and managers facing the challenging schedule environment associated with these types of hardware. It may seem a trivial matter to waive tin whisker controls for these items, but unfortunately, improper contractual specification is far too common a problem.

It has become standard procedure for certain customers to make statements such as "this is a Level 2C program," and then introduce global contractual requirements along these lines neglecting to consider the very important supporting hardware for which this level may be completely inappropriate. Subcontractors are well advised to carefully review tin control-level requirements imposed from earlier and seek early clarification with regard to supporting hardware.

6.3.2 Levels 2B and 2C – When Can Tin Be Used?

Control Levels 2B and 2C require an application-by-application consideration for every tin component in the system. Acceptability for each application will depend upon the risk posed, which will be affected by the combination of mitigations employed. As a typical system will involve many thousands of potential tin applications, a standardized methodology will need to be employed in a systematic manner.

The most commonly used approach is to define a set of design rules specifying the conditions under which the use of tin will be permitted or excluded. The details of these design rules will need to be tailored to the type of system or systems being designed and built and the applicable control level or levels.

SAE GEIA-STD-0005-2 provides rule sets for use in Level 2B and Level 2C to determine the acceptability of tin applications. These rules are summarized in the following section.

An alternative approach to the use of written design rules is the use of a risk assessment algorithm. One such an algorithm has been published and has been employed by a variety of users across the aerospace and defense industry since 2004. Some of the advantages of this approach include its ability to be applied to new classes of applications, in that it provides a standardized methodology for the collection of application-specific data. The details of this algorithm are also presented in the following section.

6.3.3   Level 3

Under Level 3 controls, the use of tin is absolutely forbidden. Given the present ubiquity of tin finishes on components, the practicality of such a regime is severely limited, as every instance of tin must be treated as a deviation from the system requirements. Use of this level is generally restricted to space-based, strategic missile, and other similar maximum-cost, maximum-consequence applications.

The principal activities for satisfying the requirements of this level involve the extreme vigilance required to prevent the leakage of tin into the system. There are two basic approaches to take: the receipt of assurances from your suppliers that they will not deliver you tin and inspection of tin finishes that you receive. Anecdotal evidence from the inspection of tin finishes received by aerospace and defense OEMs indicates that approximately 1% of all components received with documentation stating that they are not tin are, in fact, tin. Therefore, there is no real alternative to performing inspection of received components to achieve complete exclusion of tin.

6.3.4   Risk Management Techniques

Three approaches are used when making decisions on an application-by-application basis as to whether the combination of mitigating and risk factors results in an appropriate level of risk.

The first is to develop a set of design rules for use in describing applications that will be permissible or will be excluded. Most designers of commercial electronics use this approach. Standard SAE GEIA-STD-0005-2 provides guidance on the selection of design rules deemed to be acceptable for control Levels 2B and 2C for use by aerospace and military manufacturers. The following is detailed guidance on required mitigations.

Level 2B – mitigation requirements per GEIA-STD-0005-2A; any one mitigation is acceptable:

- Hard potting or encapsulation
- Physical barriers
- Circuit design and analysis showing low impact of tin whisker short or Foreign Object Debris (FOD)
- Circuit design and analysis showing that areas sensitive to tin whisker shorts or FOD have at least a 1 cm gap
- Parylene conformal coating with validated coverage and gap size, prior to coating, greater than or equal to 150 μm (5.9 mils)
- Other, nonparylene, conformal coating with validated coverage and gap size, prior to coating, greater than or equal to than 250 μm (9.8 mils)
- Pb-free tin electronic components with gaps greater than 2000 μm (78.7 mils) that have been installed with SnPb and are physically isolated from any Pb-free tin mechanical piece parts

- SnPb soldering process with validated complete coverage
- Mitigation or combination of mitigations approved by the customer.

Level 2C – mitigation requirements per SAE GEIA-STD-0005-2A; any one mitigation is acceptable.

- Hard potting or encapsulation
- Physical barriers
- Circuit design and analysis showing low impact of tin whisker short or FOD
- SnPb soldering process with validated complete coverage
- Parylene conformal coat with validated coverage and gap, prior to coating, greater than or equal to 250 μm (9.8 mils)
- Other, nonparylene, conformal coat with validated coverage and gap, prior to coating, greater than or equal to 500 μm (19.7 mils)
- A combination of mitigations approved by the customer.

The second approach is to utilize Monte Carlo techniques to estimate the failure rate associated with a particular application of tin, then comparing the predicted failure rate against the set of requirements to determine whether the application will be acceptable. A special purpose software tool has been developed by the University of Maryland for this purpose, although this technique has been used by various investigators. The benefit of this approach is that it provides a quantitative statement of risk as a function of time, which can be very useful in many circumstances. However, a limitation of this approach is that the accuracy of the prediction depends upon the similarity between the whiskering propensity of the tin in question and the tin that was used to generate the probability distribution functions used in the Monte Carlo simulation.

Since the whiskering propensity of tin finishes is known to exhibit very high degree of variability, this creates significant levels of uncertainty in the reliability of such results. The performance of Monte Carlo simulations for individual applications of tin requires much more effort and expertise than does the application of written design rules. Therefore, the use of such techniques is typically limited to applications of unusual importance or for use in justifying easier to implement written design rules.

A third approach that has been used is to employ an expert system algorithm. Such tools have been developed and used across the industry as a means to define semi-quantitative design rules for tin usage. The details of the most widely used algorithm are presented next. The tool outputs a score in the range between 0 and 10. This algorithm has been calibrated in two different ways. The yes/no decisions result from its application and have been compared against the opinions of selected subject matter experts, for threshold values of 7.0 and 7.5. The tool has also been used to assess every published tin whisker failure experience that includes sufficient detail to make an assessment. Every published tin whisker failure experience has resulted in a score of 8.95 or higher.

The intent of the algorithm is to assess the risk that for a given application of tin plating, tin whiskers would bridge between conductors. The term "overall mechanical risk" is used to describe this risk of whisker bridging. This algorithm does not address the consequences of the formation of such a bridge (electrical risk). Experience indicates that for many applications to be assessed, the risk of a whisker bridging is so negligible that further assessment of the consequences is unnecessary. Experience has also shown that in a sizable fraction of the assessments where the mechanical risk is high, the consequences of a bridge are so evident that no further risk assessment needs to be performed. For these reasons, it is anticipated that the vast majority of risk assessments need to only consist of the evaluation of the mechanical risk.

The approach taken in formulation of the algorithm is that the mechanical risk is a product of the probability that whiskers will form and the probability of these whiskers bridging between conductors. The factors that affect whisker growth relate to the properties of the plating and substrate onto which it is plated. The factors that affect the bridging risk relate to geometry of the assembly and the presence or absence of insulating coatings on the conductors. It should be noted that this algorithm is based upon the premise that failure only occurs if a whisker bridges the entire gap between conductors. This premise applies to most applications, but not to high-voltage applications where arcing across gaps is a common failure mode. Therefore, this algorithm may produce misleading results if applied to assess applications where high voltages are present.

The output of the algorithm is a numerical index of relative risk of whisker bridging, and as the levels of risk are anticipated to range over several orders of magnitude, the numerical index will be reported on a log-10 scale. Scaling factors have been selected so that the range of the numerical factor falls between 0 and 10. Higher output numbers indicated higher degrees of risk.

There are 13 inputs used for the algorithm, which represent risk and mitigation factors that affect the probability of the formation of a whisker bridging between adjacent conductors. These factors are defined as follows.

$r1 = f1$(conductor spacing)

$r2 = f2$(Pb content in plating)

$r3 = f3$(Sn deposition process)

$r4 = f4$(Sn deposit thickness)

$r5 = f5$(composition of material directly beneath the Sn deposit)

$r6 = f6$(substrate controlling the CTE imposed on the Sn deposit)

$r7 = f7$(reflow of Sn deposit)

$r8a = f8a$(type of conformal coating applied directly over the Sn deposit)

$r8b = f8b$(type of conformal coating applied on the surface of the adjacent conductors)

$r9 = f9$(use of mechanical hardware that applies stress to the surface of the Sn deposit) r10 = f10(vulnerability of the assembly to contamination related failure, as indicated by imposed environmental controls during assembly)

r11 = f11(use of conformal coating on conductors throughout the assembly)

r12 = f12(airflow within the assembly)

r13 = f13(surface area of the tin on the component)

The functions fx are as defined by Table 6.3, and the values have been adjusted during the calibration process for the algorithm.

The scale factor has been set to $K = 8.9$, based upon the maximum and minimum values produced by the functions defined as follows, to set the range of the numerical output to range from 0 to 10.

These factors are combined in accordance with the following logic.

$$\text{Overall mechanical risk} = R_{\text{total}}$$

$$\text{Total susceptibility risk factor} = R_{\text{susceptibility}}$$

This represents the effects of geometry on the ability of a whisker to create a bridge.

$$\text{Overall whisker growth risk factor} = R_{\text{formation}}$$

This represents the risk of forming a sufficient length of whisker to create a bridge.

Scaling constant $= K$

$$R_{\text{total}} = K + \log 10(R_{\text{susceptibility}} \times R_{\text{formation}}) \qquad (6.13)$$

The susceptibility of the application to whisker-induced failures is broken into two parts: primary shorts and secondary shorts. Primary shorts occur when a whisker bridges directly from its origin to an adjacent conductor. Secondary shorts occur when whiskers become dislodged and migrate through the system to a remote site with a bridge between two other conductors. The formation factor is also broken into two parts: the density of the whisker growth and the lengths of the whiskers.

$$R_{\text{total}} = K + \log 10[(R_{\text{primary}} \times R_{\text{secondary}})(R_{\text{density}} \times R_{\text{length}})] \qquad (6.14)$$

A simplification is made to formulate the risk that whiskers will grow by assuming that there are four, independent driving mechanisms of concern:

1. Stress induced during initial tin deposition
2. Stress developed in the tin as a result of interdiffusion with the material below during time/temperature exposure
3. Stress developed over time due to differential CTE between the tin and the controlling substrate

TABLE 6.3 Selection of Values of the 13 Factors for Use in the Algorithm

Risk Factor						
Conductor gap (mm)	1.27	1.0–2.0	2.0–3.0	>3.0		
Pb content (wt%)	<0.2	0.2–1				
r(2)	1	0.2	0.1	0.01	0.0001	
Process	Electroplate	Immersion		Hot dip		
r(3)	1	0.3		0.1		
Tin thickness (microns)	<1.25	1.25–6.5	6.5–13	13–25	>25	
r(4)	0.7	1	0.7	0.3	0.1	
Material directly beneath tin	Brass/bronze/BeCu	Copper	Ferrous	Nickel	Other	
r(5)	1	0.7	0.5	0.2	0.5	
Substrate controlling CTE	Ceramic	Low expansion alloy	Cu or Al	Ferrous	Other	
r(6)	1	2.5	0.2	0.3	0.5	
Plating heated after deposition	No	Annealed	Fused			
r(7)	1	0.5	0.2			
Conformal coat	None	Urethane > 1 mil	Silicone > 1 mil	Parylene	Acrylic	Other
Directly on tin surface – r(8a)	1	0.15	0.25	0.10	0.45	0.15
On adjacent conductors – r(8b)	1	0.05	0.05	0.01	0.05	0.5
Use of Mechanical HWD	Fasteners compressed onto surface	None				
r(9)	1	0.1				
Where was assembly performed	Clean room	Special clean area	Typical factory	Field assembly		
r(10)	1	0.5	0.2	0.1		
Exposed shortening sites within enclosure	Many	Some	Few	Almost none	None	
r(11)	1		0.4	0.1	0.01	
Airflow within assembly	Forced air		Dynamic use		None	
r(12)	1	0.7	0.5	0.1		
Size of component	Large mechanical item	Small mechanical item	Wire or contact	Multilead electronic	Two-lead electronic	
r(13)	2	0.25	0.1	0.01	0.001	

Note: Working versions of this algorithm in spreadsheet format are available online at various web sites.

4. Stress induced as a result of externally applied forces.

Initial stress risk factor $= R_i$

Diffusion stress risk factor $= R_d$

CTE stress risk factor $= R_{CTE}$

External risk factor $= R_{ex}$

The growth of whiskers across the gap will be diminished by the presence of conformal coating directly on the tin surface. Therefore, the four factors identifying sources of stress are combined with a conformal coat factor defining the overall risk of whisker growth.

$$R_{density} = r8a(R_i + R_d + R_{CTE} + R_{ex}) \qquad (6.15)$$

Investigations into the distribution of whisker lengths that grow from various deposits of tin indicate that some mitigation techniques are effective not because they necessarily decrease the density of whisker growths, but because they seem to restrict the lengths of the whiskers that do form. Therefore, the length factor is defined as a function of the individual factors representing plating process, substrate composition, and post plate heat treatment as follows:

$$R_{length} = r3r5r7 \qquad (6.16)$$

Combing Equations 6.13–6.16

$$R_{total} = K + \log 10((R_{primary} \times R_{secondary})\{(r3r5r7)[r8a(R_i + R_d + R_{CTE} + R_{ex})]\}) \qquad (6.17)$$

Each of the six Rx remaining values in Equation 6.17 is calculated based upon attributes of the application.

$$R_{primary} = f\{r1, r8b\}$$

$$R_{secondary} = g(R_{length}, r10, r11, r12, r13)$$

$$R_i = h\{r2, r3, r4, r5, r7\}$$

$$R_d = l\{r2, r5, r7\}$$

$$R_{CTE} = m\{r2, r6\}$$

$$R_{ex} = n\{r2, r9\}$$

Functions f, g, h, l, m, and n are functions. These functions are simple products. These functions could be redefined later if data indicates a different type of relationship applies.

The values for the 13 input factors as a function of application data are defined in Table 6.3.

The values listed on the selection table have been adjusted for the calibration process. This calibration process involves two different steps. Firstly, the widest possible selection of documented tin whisker failures was analyzed using the algorithm to check whether or not these cases would have resulted in uniformly high scores. Secondly, a wide range of actual applications of tin were assessed, and in parallel, a range of subject matter experts (SMEs) who had been routinely performing tin whisker risk assessments were asked to provide their opinion on the suitability of tin in each of these applications.

The values on the chart were adjusted so that the algorithm would faithfully reproduce the opinion of the SMEs with regard to suitability, while still providing uniformly high scores for the documented failures.

The net result of the calibration process is that all of the documented failures yield a score of 8.95 or higher, while applications where the SMEs generally agreed that tin was suitable for use scored below the range of 7.0–7.5.

These scores are typically compared against the threshold value that is agreed upon as appropriate for the reliability requirements of the system in question. In the context of system-level controls in accordance with GEIA-STD-0005-2, a threshold value of 7.5 is recommended for use with Tin Control Level 2B, and a threshold value of 7.0 is recommended for use with Tin Control Level 2C.

6.3.5 Tin Detection

The most commonly employed inspection technique for the identification of surface finish composition (and thickness) is X-ray fluorescence (XRF). As with any analytical technique, XRF has limitations that must be accounted for when it is used as part of a meaningful inspection regime. The most important limitation is that it can only be used to detect tin on external surfaces. Pure tin on the interior of components with internal cavities will not be detectable unless the parts are opened, which typically results in their destruction. (It should be noted here that many highly visible tin-whisker-induced failures in the literature were the result of whisker growth in such internal cavities.) For such devices, it will be necessary to perform some level of destructive analysis or to perform inspections at the component supplier, prior to sealing of the device.

The use of XRF has grown significantly since the introduction of the RoHS regulation, as many suppliers of commercial product require means to validate the exclusion of lead. This has benefited those who need to assure the presence of lead in tin-based finishes as the availability of a wide range of increasingly capable XRF devices has expanded in recent years. However, a device optimized for assuring the absence of lead is not necessarily optimized for assuring the presence of lead in a particular location.

For assuring tin whisker risk mitigation, it is necessary to determine that there is lead alloyed into tin above an acceptable minimum (typically 3% by weight) directly on a termination. This requires that the region being measured, referred to as the "spot size," is small enough so that meaningful measurements can be taken from the surface of a single termination. The ideal device would have a very small spot size capable of being stably placed on a single termination. Otherwise, lead that is present in other locations, such as solder, other terminations, within glass, etc., may be detected and result in an erroneous finding that the termination in question contains lead when it does not. When making measurements to assure that an assembly contains no lead, a larger spot size may often be preferable, since the detection of land from all surfaces can be important. Requirements for performing XRF measurements of Pb in tin can be found in JEDEC JESD213 [22] and Mil-STD-1580, Requirement 9 [23].

A good example of a device that is better suited for RoHS screening than for screening out pure tin parts is the handheld XRF. These devices are highly portable and resemble a radar gun. Their portability and ease of use afford many advantages including the ability to screen parts contained in bins in storage for RoHS compliance. However, their positional instability and typically large spot size are not optimum for reliable exclusion of pure tin.

An alternative technique that is often used either alone or in conjunction with XRF is energy dispersive X-ray spectroscopy (EDS) as part of a scanning electron microscope (SEM). The principal limitation of this technique is that the component in question needs to be introduced into the high-vacuum system of the SEM, which limits the size of components that can be measured as well as throughput.

Any measurement technique that is applied must account for the inherent complexities involved in the tin–lead system. Tin and lead form a eutectic system that exhibit very limited solid solubility between constituents. As a consequence, all tin–lead alloys consist of two phases: a lead-rich phase and a tin-rich phase. The relative size of these phases varies depending upon the process history and tends to coarsen with time. Therefore, obtaining a meaningful measurement of the average alloy composition requires that the spot size be sufficiently large to cover a representative number of both phases. This can be a particular problem for SEM/EDS where the effective spot size can be quite small, on the order of 1 μm, which typically permits measurement to be taken on a single phase. It is also highly recommended that compositional standards be used in the calibration of any measurement device. Compositional standards for lead in tin that are traceable to NIST standards have recently become commercially available.

6.4 CONTROL OF SUBCONTRACTORS AND SUPPLIERS

The overwhelming majority of pure tin finishes are introduced into high-reliability systems through purchased components, assemblies, and subsystems. Therefore, a crucial element of tin whisker risk management is the control over purchased items. In principle, this is no different than the control over any of the other critical attributes of purchased items. The types of controls that are practical and useful will vary depending upon the type of supplier and the type of relationship between the supplier and the customer.

Appropriate supply chain management will vary depending upon the willingness of the supplier to implement special tin controls and also the level of complexity of the purchased item, that is, components, simple assemblies, major subassemblies, etc.

6.4.1   Assemblies and Major Subassemblies

Difficulties arise when there are significant differences between the requirements of the supplier's primary customer base and your specific company requirements. With respect to pure tin, a disconnect occurs when the supplier's principal customer base is building systems that must comply with the RoHS regulations. This inevitably leads suppliers to prefer the use of pure tin over other finishes. When purchasing from such a "true COTS" supplier, it is usually a waste of time to attempt to impose detailed requirements regarding surface finishes. In many cases, the supplier will simply ignore your requirements and ship you their standard pure-tin product without taking notice of your special requests. When the ultimate customer complains that this tin violates your contractual requirements, the fact that you have imposed impractical tin requirements on your noncompliant supplier will not provide any comfort. A far better approach is to simply to assume that the COTS hardware will be utilizing pure-tin finishes and plan accordingly.

In instances where the supplier or subcontractor specializes in providing hardware to the high-reliability market, it may be possible to devolve some of the responsibility for managing tin risks onto the suppliers or subcontractors. The recommended mechanism for imposing tin whisker risk management onto competent subcontractors is the use of the SAE GEIA-STD-0005-2 standard. The simplest approach is to flow down the same control-level requirements that are being used at the next higher level of assembly. However, the actual control-level required may vary depending upon the criticality of the function, the existence of redundancy, etc. The standard requires that the supplier prepare their own tin whisker risk mitigation plan. It is recommended that customers reserve the right to review and approve such plans. With this approach, the customer needs to only determine that the plan is adequate and verify that the plan is properly executed.

6.4.2   Components

Purchasers of standard components can rarely influence their construction, so the imposition of special tin restrictions is typically not practical. However, the overwhelming majority of electronic component suppliers will provide reasonably up-to-date data on the surface finishes used on their external terminations. However, changes in construction are not uncommon and are not always communicated in a timely fashion through product change notifications. For this reason, some level of verification of termination finish by the purchaser is recommended for control Levels 2B and higher. Also, the component engineer selecting these components must understand the basic construction so that the possibility of tin in an internal cavity can be flagged when necessary.

Custom-designed components offer the systems integrator greater opportunity to control construction, but typically at a much higher cost.

The need to control the finishes on mechanical components that form part of the electronic assembly cannot be overemphasized, as these items typically exhibit much larger surface areas than do electronic components and have been identified as sources of tin whisker failures in many cases.

Some component suppliers publish data from tin whisker propensity testing, usually performed in accordance with JEDEC JESD 201 [1]. The introduction to this test standard specifically states that the results of these tests are appropriate for Class I and Class II type assemblies as defined per ANSI/J-STD-001 standard and that Class III assemblies should not utilize tin. This creates an issue for users that fall under Class III in terms of what use, if any, can be made of these test results. On the one hand, the standard explicitly excludes this type of hardware from the scope, the acceleration models behind the testing have only recently been proposed, and it is clear that the test conditions do not represent all possible driving forces for tin whisker formation. On the other hand, suppliers that actually examine some of their product after extensive and lengthy testing may be more likely to avoid serious process anomalies that could result in unusually high whisker propensity.

It is recommended that Class III users account for the fact that a component finish has passed this testing but that this factor alone is not always decisive. Within the context of SAE GEIA-STD-0005-2, employment of this type of component-level mitigation is encouraged but does not alleviate the requirement to apply one of the assembly-level mitigations listed earlier. When employing the application-specific tin whisker risk assessment algorithm discussed, components that have passed JEDEC JESD 201 [1] testing can use a lower number for one of the factors.

6.5 CONCLUSIONS

Effective and efficient tin whisker risk mitigation requires each tier of the supply chain to play its proper role. Resources are available to assist practitioners in the selection and application of effective tin whisker mitigations, which are appropriate for their products and their place in the supply chain. Industry standard SAE GEIA-STD-0005-2 provides top-level guidance on appropriate levels of controls for different classes of products and provides a standard means to communicate requirements down through the supply chain. Various specific mitigation techniques are indicated by a variety of standards. Publically available risk assessment schemes can be used to determine which mitigations are sufficient for a given product application. Standard inspection techniques are available for use in the detection of tin finishes. Taken together, proper use of this risk mitigation toolbox can be used to reduce the incidence of tin-whisker-induced unreliability to an acceptably low level.

REFERENCES

1. JEDEC Standard JESD201. Environmental acceptance requirements for tin whisker susceptibility of tin and tin alloy surface finishes. JEDEC Solid State Technology Association; 2006 March; Arlington, VA, USA.
2. Arnold SM. The growth of metal whiskers on electrical components. Proceedings of the IEEE Electric Components Conference; 1959. pp 75–82.
3. Schetty R. Minimization of tin whisker formation for lead-free electronics finishing. Circuit World 2001;27(2):17–20.
4. Moon K, Williams M, Johnson C, Stafford G, Handwerker C, Boettinger W. Proceedings of the Fourth Pacific Rim Conference on Advanced Materials and Processing. Honolulu, Dec. 11–15, The Japanese Institute of Metals, Sendai, Japan, 2001, Hanada S, Zhong Z, Nam SW, Wright RN, editors, pp. 1115–1118, 2001.

5. Glazunova VK, Kudryavtsev NT. An investigation of the conditions of spontaneous growth of filiform crystals on electrolytic coatings. J Appl Chem USSR (translated from Zhurnal Prikladnoi Khimii) March 1963;36(3):519–525.

6. Britton SC, Clarke M. Effects of diffusion from brass substrates into electrodeposited tin coatings on corrosion resistance and whisker growth. 6th International Metal Finishing Conference; 1964 May 25–29; London: Kensington Palace Hotel; 1964.

7. Endo M, Higuchi S, Tokuda Y, Sakabe Y. Elimination of whisker growth on tin plated electrodes. Proceedings of the 23rd International Symposium for Testing and Failure Analysis; 1997 October 27–31; pp 305–311.

8. Selcuker A, Johnson M. Microstructural characterization of electrodeposited tin layer in relation to whisker growth. Capacitor and Resistor Technology Symposium Proceedings; 1990 October; pp 19–22.

9. Hada Y, Marikawa O, Togami H. Study of tin whiskers on electromagnetic relay parts. Presented at the 26th Annual Relay Conference at Oklahoma State University; 1978 April; Stillwater, Oklahoma; 1978. pp 25–26.

10. http://www.lsi.com/about_lsi/corporate_responsibility/ehs/lead_free/leadfree_faqs/index .html

11. http://www.national.com/analog/packaging/tin_whiskers

12. http://www.maxim-ic.com/emmi/tin_whisker_data.cfm

13. Lee B-Z, Lee DN. Spontaneous growth mechanism of tin whiskers. Acta Mater 1998;46(10):3701–3714.

14. Fukuda Y, Osterman M, Pecht M. The effect of annealing on tin whisker growth. IEEE Trans Electron Packag Manuf 2006;29(4):252–258.

15. Fujimura I. Internal communications to CALCE.

16. Meschter S, Snugovsky P, Kennedy J, Bagheri Z, Kosiba E. SERDP tin whisker testing and modeling: High temperature/high humidity conditions. International Conference on Soldering & Reliability 2013 Proceedings, SMTA.

17. Hester T. Tin whiskers self-mitigation in surface mount components attached with leaded (Pb-containing) solder alloys. 7th International Symposium on Tin Whiskers; 2013 November.

18. Kadesch JS, Leidecker H. Effects of conformal coat on tin whisker growth. Proceedings of IMAPS Nordic, The 37th IMAPS Nordic Annual Conference; 2000 September 10–13; pp 108–116.

19. Brusse J, Ewell G, Siplon J. Tin Whiskers: Attributes and Mitigation, Capacitor and Resistor Technology Symposium (CARTS); 2002 March; pp 67–80, 25–29.

20. Hilty RD, Gorman N. Tin whisker reliability assessment by Monte Carlo simulation. IPC/JEDEC Lead-Free Symposium, San Jose, 2005.

21. Fang T, Osterman M, Pecht M. Statistical analysis of tin whisker growth. Microelectron Reliab 2006;46:846–849.

22. JEDEC JESD213 standard. Standard Test Method Utilizing X-Ray Fluorescence (XRF) for Analyzing Component Finishes and Solder Alloys to determine Tin(Sn)-Lead(Pb) Content; 2010.

23. MIL-STD-1580. Department of Defense Test Method Standard: Destructive Physical Analysis for Electronic, Electromagnetic and Electromechanical (EEE) Parts.

7

QUANTITATIVE ASSESSMENT OF STRESS RELAXATION IN TIN FILMS BY THE FORMATION OF WHISKERS, HILLOCKS, AND OTHER SURFACE DEFECTS

NICHOLAS G. CLORE

School of Materials Engineering, Purdue University, West Lafayette, Indiana, USA

DENNIS D. FRITZ

Science Applications International Corporation (SAIC), McLean, Virginia, USA

WEI-HSUN CHEN

School of Materials Engineering, Purdue University, West Lafayette, Indiana, USA; Cymer, Inc., San Diego, California, USA

MAUREEN E. WILLIAMS

National Institute of Standards and Technology (NIST), Gaithersburg, Maryland, USA

JOHN E. BLENDELL AND CAROL A. HANDWERKER

School of Materials Engineering, Purdue University, West Lafayette, Indiana, USA

7.1 INTRODUCTION

This chapter focuses on development and validation of a surface-defect counting procedure, which can be applied in research on the specific mechanisms responsible

Mitigating Tin Whisker Risks: Theory and Practice, First Edition.
Edited by Takahiko Kato, Carol A. Handwerker, and Jasbir Bath.
© 2016 John Wiley & Sons, Inc. Published 2016 by John Wiley & Sons, Inc.
Companion website: www.wiley.com/go/Kato/TinWhiskerRisks

for stress relaxation and tin (Sn) whisker formation in tin films. The current Joint Electron Device Engineering Council (JEDEC) standards [1, 2] for the evaluation of Sn whiskers are focused on identifying high-aspect-ratio whiskers and limit analysis to areas of high whisker density. The JEDEC counting procedure was, therefore, designed to identify areas of high shorting risk. Although the risk assessment of shorting by tin whiskers is important, an understanding of whisker growth mechanisms and competing mechanisms for stress relaxation requires quantitative, comprehensive data on how stress relaxation is being accommodated overall, not just by high-aspect-ratio whiskers. Furthermore, armed with that data and the resulting knowledge of mechanisms for relaxing stress other than by whisker formation, it is hoped that Sn-based microstructures can one day be designed that are resistant to whisker formation, essentially eliminating the risk of circuit shorting.

By looking at tin film surfaces, it is clear that stress relaxation occurs not only by the formation of whiskers but also by the formation of other surface defects, including but not limited to hillocks, holes, sunken grains, irregularly shaped structures, and low-aspect-ratio grains that protrude from the film surface, referred to as "popped grains." If these surface defects form to relax stresses, tin whisker formation and growth are expected to be affected. The question is how to characterize the differences. In this chapter, we describe (i) how to quantify the stress relaxation that can be attributed to these separate types of defects, each with a lower reliability risk than tin whiskers; (ii) how the data can be combined to quantify overall stress relaxation in a given film under a given set of conditions; and (iii) what comparisons of such data from the same films under different conditions, or different films under the same stressing conditions, reveal about the relationship between film microstructure, stressing conditions, and stress relaxation.

A methodology to obtain statistically significant data sets on defect formation was developed to allow for such quantitative comparisons as a function of film characteristics, manufacturing variables, sources of stress, and storage conditions. Using this methodology, differences in defect morphologies, defect dimensions, defect densities, and the total relaxed volume/film area and total relaxed volume/film volume (i.e., volumetric strain) can be assessed. Of particular interest is the comparison of the total defect volumes and the corresponding volumetric strains relaxed by defect formation.

The defect assessment methodology was successfully demonstrated with a suite of 18 sample types supplied by the International Electronics Manufacturing Initiative (iNEMI) Tin Whisker Assessment Project (see Refs [3–7] for representative iNEMI whisker study reports). Matte, satin, and bright tin surface finishes were plated on brass and tungsten substrates and were stored in ambient, thermally cycled, or exposed to constant temperature/moderate humidity conditions. The methodology documented clear differences in evolution of film microstructure and defect formation among the 18 different sample types, providing a method to quantify the complexity of stress relaxation in tin surface finishes, which are presented in detail in this chapter.

7.2 SURFACE-DEFECT CLASSIFICATION AND MEASUREMENT METHOD

In order to analyze the strain relaxed in tin coatings by the formation of surface defects, the number and volume of whiskers and other surface defects must be determined. The classes of surface defects are defined as follows:

- Surface defects with a length to diameter aspect ratio of 2 or greater and no lateral growth/grain boundary migration in the plane of the film are classified as whiskers (Figure 7.1a).
- Hillocks are surface defects that grow out of the plane of the film, as well as in diameter as the out-of-plane growth proceeds. The radial, in-plane grain growth leads to the observed increase in diameter and a characteristic change in defect shape (Figure 7.1b).
- Popped grains are entire grains that grow out of the plane of the film, with an aspect ratio of less than 2 (Figure 7.1c).
- Sunken grains are the inverse of popped grains: they are entire grains that, as a result of stress relaxation, are depressed relative to the free surface (Figure 7.1d).
- Irregular growths are grains with multiple segments, each with varying cross section, that grow out of the plane of the film. Irregular growths have been referred to elsewhere as "odd-shaped eruptions" (Figure 7.1e).

The JEDEC tin whisker counting method accounts for only the effective length of tin whiskers, that is, total vertical distance out of the plane of the film, as an indicator of the potential for electrical shorts during the operation of electronic equipment. A schematic of an effective length measurement from the JEDEC Standard JESD22-A121A [1] is shown in Figure 7.2a. The effective length is therefore an underestimate of the actual whisker length.

For the methodology presented here to quantify the relaxation provided by an individual surface defect, the defect dimensions and volume must be accurately assessed. For example, the volume of the kinked whisker in Figure 7.2b is the sum of the volumes of the segments. To accurately measure the defect length, two images of each are required at different stage tilts to account for the actual length of the defect rather than using the projected length. Using measured defect projection lengths from each image, the actual length of the defect segment can be calculated using the known stage tilt and a simple rotational matrix [8]. Figure 7.3 shows a whisker viewed normal to the film surface and at a 25° stage tilt. The apparent length is different because the whisker is three dimensional and each image is a projection of the whisker length from the viewing plane.

The volume of other surface defects is calculated from scanning electron microscope (SEM) measurements analyzed using "Image J" software [9] as follows:

- Tin whiskers – assumed to be cylindrical. Volume calculated from the cross-sectional area of a whisker multiplied by the sum of the segment lengths.

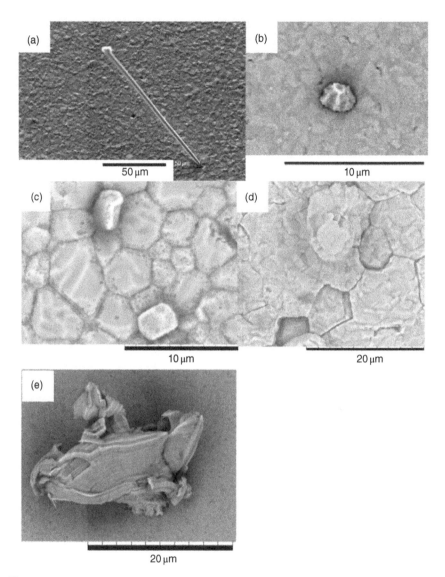

Figure 7.1 Scanning electron microscope (SEM) microstructures of surface defects using backscattered electron imaging. (a) Whisker, (b) hillock, (c) popped grains, (d) sunken grains and a popped grain, and (e) irregular growth.

- Hillock – assumed to be conical. Volume calculated from one-third of the base area multiplied by the height.
- Popped grains – volume calculated from the cross-sectional area of a popped grain multiplied by the height taken from the stage tilt procedure used for afore-mentioned whisker length.

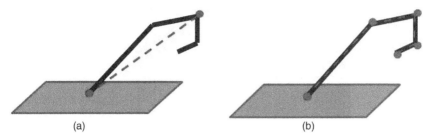

(a)   (b)

Figure 7.2 (a) Schematic of whisker length measurement technique from the 2008 version of JESD22-A121A [1]. Length of a whisker is measured using a straight line from the origin to the furthest point of the whisker. (b) Schematic of measurement of a whisker using the segmented length technique.Source: Modified schematic from [1].

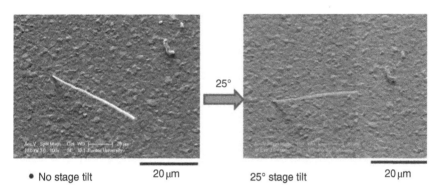

● No stage tilt   20 μm   25° stage tilt   20 μm

Figure 7.3 Comparison of a whisker observed using SEM (a) with no stage tilt and (b) with 25° stage tilt.

- Sunken grains – volume calculated from the area of grain on the surface from "Image J" software multiplied by a representative grain depth from focused ion-beam (FIB) milling cross-sectional analysis of representative sunken grains in the sample.
- Irregular growths (odd-shaped eruptions) – volume calculated based on measurements of rational segments and approximated with simple geometric shapes.

The observation and sampling procedure for this study is directed toward finding representative areas for quantifying surface defects. This contrasts to the JEDEC standard JESD22-A121A procedure [1] that examines only the area most likely to form tin whiskers and then measures the lengths at fixed time intervals – simulating whisker growth in the expected life of the tested sample. Also, JESD22-A121A [1] only counts whiskers longer than 10 microns and ignores any other surface defects present. Finally, the JESD22-A121A procedure [1] emphasizes observing edge areas, where tin electroplate properties may encourage the growth of long tin whiskers.

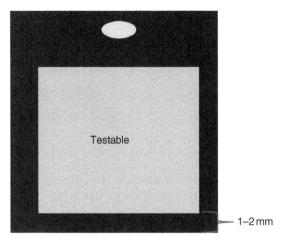

Figure 7.4   Schematic of test coupon. Testable area in gray, while the area with inconsistent tin-electroplated characteristics is shaded in black.

This study was performed on blanket, uniform tin-electroplated films, so that stress and strain are the result of test factors, not edge effects. The methodology could also be applied to samples of other geometries if the film characteristics are known as a function of position on the sample, but results become difficult to interpret. Figure 7.4 shows a schematic of the uniformly plated testable area of the test coupons used in this work. The testable areas exclude at least 1 mm of plating along any plated edge, and in some cases, film thickness measurements using microscopy on cross sections or XRF analysis may be necessary to establish the testable area.

A key step in the measurement process is to determine the total area that must be evaluated and the image magnification necessary to obtain a valid, statistically significant measurement for a given sample. For each electroplated coupon, the testable area was divided into nine rectangular regions, as indicated in Figure 7.5. To start, a measurement area of $14{,}000\,\mu m^2$ was randomly selected within each region and defined as a section, and defect densities were measured in each section. For a set of nine sections, the standard deviation and the average of the defect densities were determined in units of number of defects per section and number of defects per square millimeter. If the standard deviation was larger than the mean, a new, larger measurement area of $56{,}000\,\mu m^2$ (four times larger) was selected in each region and the defect densities were again measured at a magnification appropriate for imaging the defects. This process of increasing the measurement area while keeping the magnification of the images high enough to identify all the defects was iteratively applied until the standard deviation of the defect density was lower than the mean to reduce measurement errors due to the spatial distribution of defects while minimizing the counting time. As the defect density cannot be below zero, a large standard deviation relative to the mean implies a nonrandom spatial distribution.

Figure 7.5 Schematic of the nine regions within the testable region of a coupon.

All defects on the tin-electroplated surface are used to calculate the defect density. If a scratch or irregularity in the film surface is found in the section, another area is selected within the region. Inspections at larger areas do not have to be performed in the same area as the previous smaller area, but they remain within the same region. The measurement area for which the standard deviation of defect densities is less than the mean is then selected for the detailed inspection. At this size, the inspection area is assumed to contain a representative distribution of defects.

Table 7.1 shows an example of increasing the area for the detailed inspection, with each column representing a different sized area. To visualize the variability in defect density, a histogram is included in each column. The x-axis of the histogram is the number of defects per inspection area, ranging from 0 to 5 defects/area. The y-axis of the histogram is the number of areas with specific number of defects in the area. The summation of the values in the y-axis will always equal 9, the total number of areas inspected. The average number of defects per inspection area is listed under the histogram, as is the number of defects normalized to an area of 1 mm^2. The standard deviation of defects between sections is shown in the final rows with a normalized standard deviation to 1 mm^2 in the last row.

For an inspection area of 240,000 μm^2, the average number of defects per inspection area is 2.22 with a standard deviation of 1.79. This indicates that the detailed inspection will be conducted with a section area of 240,000 μm^2. The normalized number of defects/mm^2 shows that the initial average defect density for the smaller area was an overestimation with a large standard deviation. As the inspection area increases, the normalized average number of defects is reduced, but the standard deviation also decreases. It is expected that for an inspection over an area larger than this, performed at an appropriate image magnification, the normalized number of defects per 1 mm^2 would remain at 9 ± 7 defects/mm^2.

TABLE 7.1   Example of How to Choose Inspection Area Based on Whisker Density Statistics

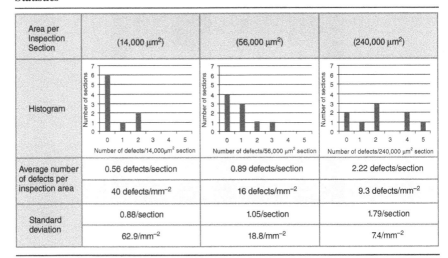

For the detailed measurements in each area, the morphology of each defect was noted and its dimensions were measured as appropriate for the defect type as defined earlier. Defect volume is calculated using the assumptions listed earlier.

Due to the spatial variability in defect density, the appropriate magnification for analysis cannot be selected in advance of the defect density measurements. In addition, uncertainty in dimensional measurements should also be established in order to allow quantifiable comparisons between samples. In the results presented next, there are open questions with respect to measurement uncertainty. In particular, the error in dimensional measurements was not quantified. Therefore, the data sets presented next are suitable for identifying trends but not for quantitative comparison.

7.3   PREPARATION AND STORAGE CONDITIONS OF ELECTROPLATED FILMS ON SUBSTRATES

As noted earlier, a suite of 18 sample types was prepared by the iNEMI Tin Whisker Assessment Project. Three different types of Sn electrolytes were used, generating three types of Sn finishes that were categorized with the common definitions of "bright," "matte," and "satin." These three finishes were electroplated on two substrates: brass and high-purity tungsten.

The brass substrates were sheared into $10\,mm \times 10\,mm$ coupons with a thickness of 0.38 mm from brass Hull cell cathodes (Kocour Company, Chicago, IL, Part No. 050062*). The composition of the brass as reported in mass% by the manufacturer was 69.325% Cu, 30.500% Zn, 0.007% Pb, and 0.038% Fe.

As-rolled tungsten sheet (99.95 mass% purity as reported by the manufacturer), 0.25 mm thick, was cut into 10 cm × 10 cm coupons by electrical-discharge machining (EDM). All of the coupons were polished to a 1 μm diamond finish, then cleaned and etched with 10% potassium hydroxide in water. The coupons were etched with an ionized argon beam for a final *in situ* cleaning of the surface. The tungsten (W) coupons required deposition of a seed layer of Sn prior to electroplating to produce an electrically conductive surface. A seed layer, 0.2 μm thick, of high-purity Sn (99.99 mass%) was vapor deposited on the substrate surfaces; a vacuum in the low 10^{-4} Pa (10^{-6} Torr) range was maintained during the deposition. This Sn seed layer ensured adherence during electroplating by preventing the formation of tungsten oxides on the substrate surface prior to and during electroplating in the methane sulfonic acid (MSA) electrolytes. The tungsten and brass substrates were plated with bright, satin, and matte Sn by the Ram-Chem* (China) facility in March 2009 using Ram-Chem* electrolytes. The electrolytes*, current densities, as-plated in-plane grain sizes, and deposit thicknesses for the three Sn surface finishes are listed in Table 7.2. Characterization techniques included electron backscatter diffraction (EBSD) and X-ray diffraction to measure crystallographic texture, and FIB cross-sectioning to characterize film thickness, grain size, and the heights of sunken grains.

Samples were randomly designated to one of three storage conditions: thermal cycling (TC), isothermal storage at a constant humidity, and ambient storage. Thermal cycling was performed following the JEDEC standard JESD201 [2], which includes 500 cycles of −45 to 85°C with a 9 min dwell at maximum and minimum temperatures. The isothermal temperature and relative humidity (RH) storage conditions were 1000 h at 50°C/50%RH. Ambient storage was at room temperature without temperature or humidity control.

7.4   SURFACE DEFECT FORMATION AS A FUNCTION OF TIN FILM TYPE, SUBSTRATE, AND STORAGE CONDITION

Before reading the following results and discussion, readers might find it useful to review sections of the chapters by Sarobol et al. (Chapter 1) [10] and Chason (Chapter 2) [11] in this book that present the fundamental models for why and how

TABLE 7.2   Film Deposition Parameters and Film Characteristics

Finish	Electrolyte[a]	Plating Current Density (mA/cm²)	Estimated As-Plated In-Plane Grain Size (μm)	Average Deposit Thickness (μm)
Bright	RamTechSnB14	108	1	5.0
Satin	RamTechSnSL28	129	3	3.2
Matte	RamTechSnM48	43	4	3.6

[a]NIST Disclaimer: Certain trade names are mentioned for experimental information only, in no case does it imply a recommendation or endorsement by NIST.

whiskers form based on microstructure, mechanisms, and driving forces and present critical experiments and tests of the various models. Of particular importance to understanding the results presented in this chapter are the discussions in Sarobol et al. [10] of the roles of crystallographic texture and microstructure on stress relaxation by whisker growth under different conditions that induce film stresses. Key points of the chapters are as follows:

1. The formation of whiskers over other forms of stress relaxation is critically dependent on the interactions between local microstructure, stress, and the kinetics of various competing processes for stress relaxation.

2. Whiskers and hillocks generally form from shallow surface grains, in accordance with the physics-based model for growth based on grain boundary sliding-limited, highly localized Coble creep. The geometry of an individual grain (in-plane diameter, depth as characterized by the grain boundary angle θ), the grain boundary sliding behavior of the grain boundaries that bound it, and the imposed stress all play a role in determining whether it can grow as a whisker.

3. For equivalent microstructures, a preferred crystallographic texture of a film may produce a high propensity for whisker formation under one condition of stress, for example, for intermetallic growth (elastic stress), but a different propensity due to thermal expansion mismatch with a substrate (thermoelastic stress), with different textures showing different effects. (Chapter 1, Figure 1.11). For example, a film with a strong (001) texture for elastic stress conditions is expected to have a high overall elastic strain energy density (ESED) and large ESED differences between grains, compared to the other textures. Under thermoelastic stress conditions, however, the same film has a low overall ESED, with small differences between grains. In contrast, a film with a strong (110) texture for elastic stress conditions has a slightly lower overall ESED than an equivalent strongly textured (001) film and large differences between grains, while for thermoelastic stress conditions, the (110) film has a high overall ESED and large differences between grains. These simulations and related findings that grains with high local ESED are more prone to whisker formation highlight both the overall effects of texture and the local conditions for whisker growth in Sn films. The thickness and strength of a surface oxide layer and its continued growth are important to whether fracture of the oxide can occur for a whisker to initiate and continue to grow.

The results are presented here in two sections, based on substrate type, with three different films and with three different storage/stress conditions. The rationale for this separation is based on the sources of stress in the films. With the brass substrates in contact with Sn films, a continuing source of stress is due to the formation of Cu–Sn intermetallic compounds at the Sn–brass interface regardless of the storage conditions. In contrast, there are no Sn–W intermetallics that form in the Sn–W system, so the sources of stress are predominantly due to thermal expansion mismatch between

Sn and W for thermal cycling and perhaps even to some extent from constant temperature/humidity annealing. Within each section, based on the substrate, an example is presented to illustrate how these detailed measurements of surface defect formation provide insights into the interactions between microstructure, storage condition/stress generation, defect type, sizes, densities, and volume relaxed. The full data sets and more information on the individual sample types can be found elsewhere [12].

7.4.1 Tin Films on Brass Substrates

Figure 7.6 shows typical microstructures of bright tin on brass (BonB), satin on brass (SonB), and matte on brass (MonB) after conditions of ambient storage, thermal cycling (TC), and constant temperature/%relative humidity (T/RH) storage. There are a few qualitative observations made regarding the effects of storage condition on microstructural changes. The average grain diameter of the BonB samples increased following thermal cycling, with noticeable void formation in the film and sunken grains. The MonB films experienced an increase in average in-plane grain diameter following thermal cycling.

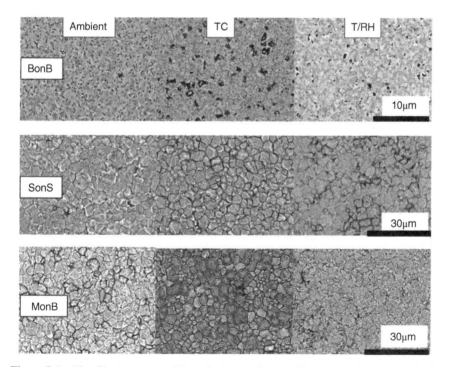

Figure 7.6   Nine film/storage conditions for brass substrates. From top to bottom: bright tin on brass (BonB), satin on brass (SonB), and matte on brass (MonB). From left to right, storage condition of 2 years at ambient, TC = 500 cycles from −45 to 85°C, T/RH = 1000 h at 50°C/50%RH.

An important film characteristic that affects how individual films respond to applied stresses or strains is a film's crystallographic texture, a measure of the arrangement of the crystallographic orientations in a polycrystalline material relative to a sample axis [13]. The crystallographic texture was determined using X-ray diffraction and/or EBSD, and from this information, two texture-dependent thermophysical properties (biaxial modulus and coefficient of thermal expansion (CTE)) were calculated assuming perfect fiber texture. Table 7.3 gives the texture axis (hkl) and texture strength as indicated by multiples of a random distribution (MRD) and the corresponding calculated biaxial modulus and CTE.

The differences in the average biaxial modulus and CTE are small for the three film types. For a stress created isothermally by IMC formation, these average modulus differences may not appear to be significant, but based on the texture research results in Chapter 1, the local variations in ESED, stress, and strain are affected by the texture and MRD. The strain differences created by the CTE mismatch between Sn films with this texture and brass ($19-21 \times 10^{-6}/°C$) lead to an approximate tensile linear elastic strain in the film during heating of approximately 0.0003 for a ΔT of 85°C, which switches to compressive on cooling assuming no stress relaxation. (Note that this is in contrast to copper with its CTE of approximately $17 \times 10^{-6}/°C$, in which the thermal expansion mismatch induced stresses would be close to zero or compressive during heating.) Tin films on brass substrates stored in ambient conditions should have stresses dominated by intermetallic growth and negligible stress due to CTE mismatch. The temperature–humidity conditions are used in the JEDEC standard to induce oxide formation and corrosion of Sn films.

A summary of all measurements for Sn films on brass is provided in Table 7.4 and Figure 7.7 in two configurations to allow for easier comparison. The top three tables compare the behavior of the three finishes with a given storage condition. The bottom three tables compare the behavior of a single finish under the three storage conditions. The total defect volume per normalized area of 1 mm^2 is converted into a volumetric strain relaxation by multiplying the normalized area by film thickness. This is equal to the volumetric strain relaxed by formation of surface defects. The underlying measurements for the contribution of different defect types to stress relaxation on which the tables are based are presented similarly in Figure 7.7. Histograms show the number and total relaxation volumes associated with each defect type for different film types with different storage conditions. All measurements in Figure 7.7 were

TABLE 7.3   Measured Texture Orientation and Calculated CTE and Biaxial Modulus for As-Electroplated Films on Brass Substrates Assuming Perfect Fiber Texture

Surface Finish	Preferred Texture (hkl)	MRD	Calculated Biaxial Modulus (GPa)	Calculated CTE ($10^{-6}/°C$)
BonB	112	22	91	18
SonB	001	20	92	18
MonB	001	35	94	17

TABLE 7.4 Summary of Defects and Volumetric Strains for Nine Film/Storage Conditions for Sn-on-Brass Substrates: (a) Sorted by Storage Condition: 2 years at Ambient, TC = 500 cycles from −45 to 85°C, T/RH = 1000 h at 50°C/50%RH and (b) Sorted by Film Type Bright (BonB), Satin (SonB), and Matte (MonB)

(a)

Surface Finish	Defect Density (number/mm^2)	Total Defect Volume (μm^3/mm^2)	Volume per Defect (μm^3/Defect)	Volumetric Strain (mm^3/mm^3)
Ambient				
BonB	22	6,185	278	1.20E−03
SonB	17	1,376	80	4.30E−04
MonB	733	30,257	41	8.40E−03
Thermal cycling				
BonB	21	25,980	1,220	5.20E−03
SonB	113	4,422	39	1.40E−03
MonB	756	28,793	38	8.00E−03
Constant temperature/% relative humidity				
BonB	10	10,768	1,077	2.20E−03
SonB	104	3,664	35	1.10E−03
MonB	233	12,503	54	3.50E−03
(b)				
Storage				
Bright on brass (BonB)				
Ambient	22	6,185	278	1.20E−03
T/RH	10	10,768	1,077	2.20E−03
TC	21	25,980	1,220	5.20E−03
Satin on brass (SonB)				
Ambient	17	1,376	80	4.30E−04
T/RH	104	3,664	35	1.10E−03
TC	113	4,422	39	1.40E−03
Matte on brass (MonB)				
Ambient	733	30,257	41	8.40E−03
T/RH	233	12,503	54	3.50E−03
TC	756	28,793	38	8.00E−03

normalized to a total sample area of 2.16 mm^2. The pie charts show the percentages of volume relaxed by each defect type. One comparison of individual whisker dimensions as a function of storage condition and film type is shown in Figure 7.8, with the other plots provided in the Chapter Appendix.

The most apparent trends in Table 7.4 and Figure 7.7 are as follows:

(a) Whisker formation was the dominant stress relaxation defect type in two of the three films (SonB and MonB) for all storage conditions and contributed

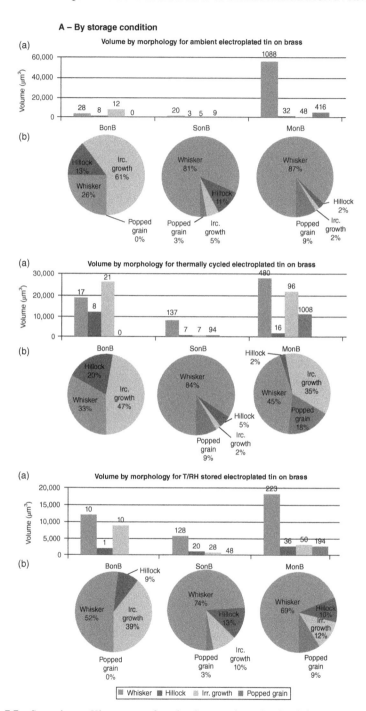

Figure 7.7   Sn on brass: Histograms of total volume and number by defect types organized by storage condition (Column A on left) and by film type (Column B on right). The same nine data sets are presented in Column A and Column B but are arranged differently to more easily compare results.

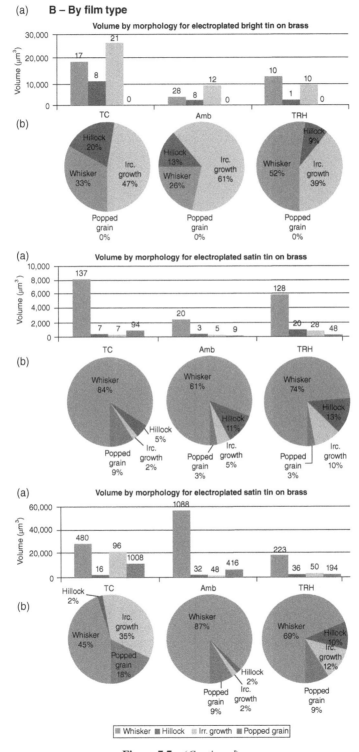

Figure 7.7 (*Continued*)

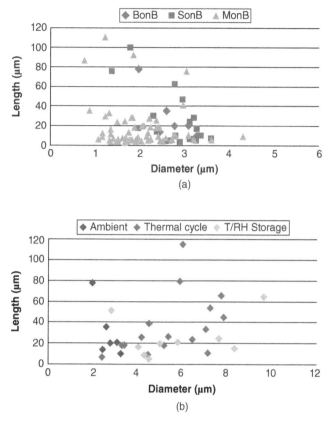

Figure 7.8 (a) Scatter plot of whisker growth on bright, satin, and matte Sn-electroplated films on brass substrates after 2 years of ambient storage. (b) Scatter plot of whisker growth on bright Sn-electroplated films on brass substrates at three different storage conditions: 2 years at ambient, TC = 500 cycles from −45 to 85°C, T/RH = 1000 h at 50°C/50%RH.

significantly to relaxation. In BonB, irregular growth volume is larger than whisker volume in two of the three storage conditions.

(b) The overall defect density is largely determined by the film type, rather than by the storage condition.

(c) By comparing the rows in Column B in Figure 7.7, the types and distributions of defects observed are also largely determined by film type, rather than storage condition; the relative volumes of each defect type within a film vary with storage condition but with no simple trend as a function simply of storage condition.

(d) The volumetric strains for thermal cycling are not the same for the three films, suggesting that defect formation is working in competition with other stress relaxation mechanisms with different kinetics for each film type.

These combinations of measurements, particularly the detailed measurements of surface defect dimensions, reveal trends in stress relaxation by surface defect formation that individual measurements are unable to provide. When taken with individual measurements of whisker dimensions, as seen in Figure 7.8, additional information about which grains form whiskers in a given film under a given set of conditions can be obtained. For example, the in-plane grain sizes of BonB, SonB, and MonB are 1, 3, and 4 μm, respectively, for the associated film thicknesses of 5 μm (BonB), 3.2 μm (SonB), and 3.6 μm (MonB). In Figure 7.8a showing ambient storage of the three film types, all three film types show a maximum whisker length that decreases with increasing whisker diameter. BonB films produce few whiskers, but all are significantly greater in diameter than the average in-plane grain size. SonB film whiskers span the range of diameters from 1.3 to 3.6 μm. In contrast, all MonB whiskers with the exception of one whisker have diameters less than the average in-plane grain size. Without more information on the cross-sectional geometries, it is not possible at this point to relate these observations to differences in subsurface grain geometries or intermetallic growth. These results do, however, illustrate the level of detailed measurement and analysis needed to separate the effects of various factors on stress relaxation in Sn films.

7.4.2   Electroplated Tin on Tungsten Substrates

Figure 7.9 shows typical microstructures of bright tin on tungsten (BonW), satin on tungsten (SonW), and matte on tungsten (MonW) after conditions of ambient storage, thermal cycling (TC), and constant temperature/%relative humidity (T/RH) storage. There are a few qualitative observations made regarding the effects of storage condition on microstructural changes. It was difficult to estimate the final average grain size after storage since the surface morphologies were dominated by the shapes of the grains in the as-plated films, which previous FIB studies have shown do not always correspond to grain boundaries. The positions of grain boundaries could be detected unequivocally when there was cracking, grain sinking, or defect formation, which did not occur for most grains. The in-plane grain size for all three film types increased significantly with thermal cycling. Thermal cycling of the three film types also resulted in increased void formation and sunken grains, that is, the other storage conditions did not produce sunken grains. The MonW T/RH sample was not analyzed due to sample damage.

Table 7.5 lists the CTE for each sample based on its texture orientation measured by EBSD. The multiples of random distribution (MRD) values indicate the strength of the given texture and are used to calculate the biaxial modulus and CTE. With the CTE for W of $4.3 \times 10^{-6}/°C$, compressive strains of approximately 0.001 are expected in the Sn films for a temperature increase of 85°C.

A summary of all measurements for Sn films on tungsten is provided in Table 7.6 and Figure 7.10 in two configurations to allow for easier comparison. The top three tables compare the behavior of the three finishes with a given storage condition. The bottom three tables compare the behavior of a single finish under the three storage conditions. The total defect volume per normalized area of $1 mm^2$ is converted into

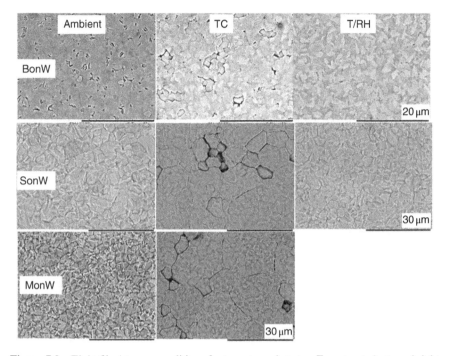

Figure 7.9   Eight film/storage conditions for tungsten substrates. From top to bottom: bright tin on tungsten (BonW), satin on tungsten (SonW), and matte on tungsten (MonW). From left to right: 2 years at ambient, TC = 500 cycles from −45 to 85°C, T/RH = 1000 h at 50°C/50%RH. Matte-on-tungsten T/RH sample was not analyzed due to sample damage.

TABLE 7.5   Measured Texture Orientation and Calculated CTE and Biaxial Modulus for Samples Electroplated on Tungsten Substrates, Assuming Perfect Fiber Texture

Surface Finish	Preferred Texture (hkl)	MUD	Calculated Bi-axial Modulus (GPa)	Calculated CTE ($10^{-6}/°C$)
BonW	001	14	91	18
SonW	001	22	93	17
MonW	001	42	95	17

a volumetric strain relaxation by multiplying the normalized area by film thickness. This is equal to the volumetric strain relaxed by formation of surface defects. The underlying measurements for the contribution of different defect types to stress relaxation on which the tables are based are presented in Figure 7.10. Histograms show the number and total relaxation volumes associated with each defect type for different film types with different storage conditions. All measurements in Figure 7.10 were normalized to a total sample area of 2.16 mm^2. The pie charts show the percentages

TABLE 7.6 **Summary of Defects and Volumetric Strains for Eight Film/Storage Conditions for Sn-on-Tungsten Substrates: (a) Sorted by Storage Condition: 2 years at Ambient, TC = 500 cycles from −45 to 85°C, and T/RH = 1000 h at 50°C/50% RH, (b) Sorted by Film Type Bright (BonB), Satin (SonB), and Matte (MonB)**

	Defect Density (mm^{-2})	Defect Volume (μm^3/mm^2)	Volume (μm^3/Defect)	Volumetric strain
(a)				
Ambient				
BonW	304	1,662	5	3.3E−04
SonW	91	904	10	2.8E−04
MonW	169	804	5	2.2E−04
Thermal cycling				
BonW	1,911	11,423	6	2.3E−03
SonW	2,457	11,622	6	3.6E−03
MonW	1,800	12,807	7	3.6E−03
Constant temperature/humidity				
BonW	815	3,990	5	8.0E−04
SonW	237	3,317	14	1.0E−03
MonW	X	X	X	X
(b)				
Bright on tungsten (BonW)				
Ambient	304	1,662	5	3.3E−04
T/H	815	3,990	5	8.0E−04
TC	1,911	11,423	6	2.3E−03
Satin on tungsten (SonW)				
Ambient	91	904	10	2.8E−04
T/H	237	3,317	14	1.0E−03
TC	2,457	11,622	6	3.6E−03
Matte on tungsten (MonW)				
Ambient	169	804	5	2.2E−04
T/H	X	X	X	X
TC	1,800	12,807	7	3.6E−03

of volume relaxed by each of the five defect types. Individual whisker dimensions as a function of storage condition and film type are provided in the Chapter Appendix.

The most apparent trends for Sn on W as seen in Table 7.6 and Figure 7.10 are as follows:

(a) Popped grains are the dominant stress relaxation feature in terms of number and total volume relaxed for all film types and conditions. This is in marked contrast to Sn on brass, where popped grains were minority defects both in number and total volume.

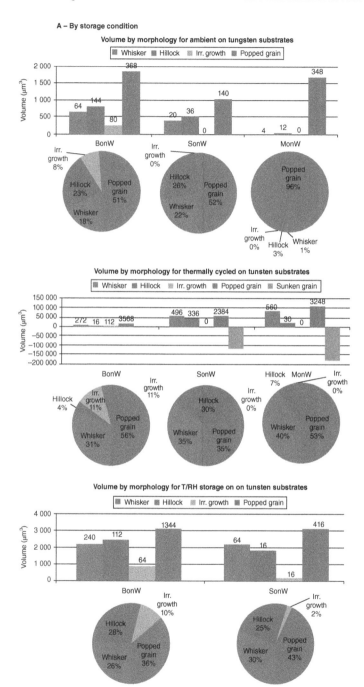

Figure 7.10   Sn on W: Histograms of total volume and number by defect types organized by storage condition (Column A on left) and by film type (Column B on right). The same eight datasets are presented in Column A and Column B but are arranged differently to more easily compare results. The number and total of sunken grains are reported in the histograms where relevant but are not included in the pie charts.

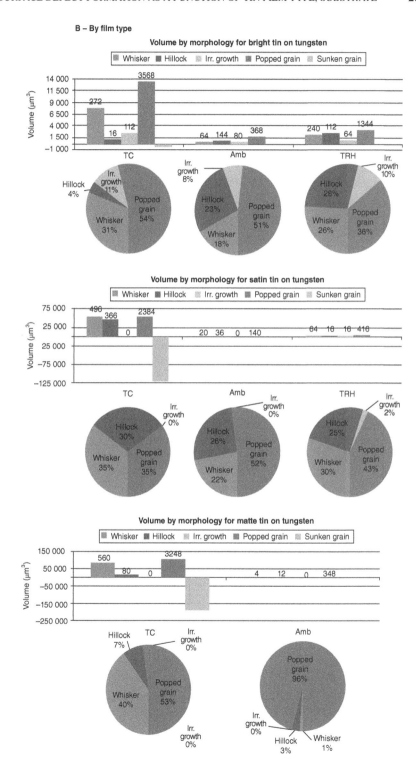

Figure 7.10   (*Continued*)

(b) The defect densities for each film type and storage condition are significantly higher for Sn on W than for their equivalent Sn on brass samples, except for MonW after ambient storage.

(c) The net volumetric strains for Sn on W are approximately the same for all three film types and the same order of magnitude as Sn on brass. However, the presence of sunken grains makes stress relaxation different for the cases of SonW and MonW, as discussed in more detail next.

All of the Sn films on W stored for 2 years under ambient conditions had defect formation even in the absence of IMC growth or significant CTE-induced stresses. Popped grains dominated the defect population, and defect formation produced a total volumetric strain of approximately 3×10^{-4}. If one assumes a 5°C temperature fluctuation during storage, this leads to a linear elastic strain of 6×10^{-5}, and therefore a volumetric strain of 2×10^{-4}, which is the same order of magnitude as observed. Residual plating stresses or stresses due to surface oxide growth may have led to the whisker and surface defect formation; however, the specific sources of stress in these films are unknown.

During thermal cycling, the films are exposed to extreme temperature excursions that cause alternating compressive and tensile stresses. The compressive stresses induced whisker growth and thousands of popped grains on all three film types. As noted in Chapter 1, tensile stresses can cause pre-existing defects, which grow under compressive stress to shrink in height or other film grains to sink below the free surface of the film, as these defect grains become sources of Sn to relax tensile stresses elsewhere in the film. Thus, the volume removed by sinking grains under tensile stress may eliminate some or all of the volume increase occurring under compressive stress, leading to a net volume that could be positive or negative depending on overall stress relaxation processes. This situation occurred for SonW and MonW during thermal cycling, with the volumes of each type of defect (whiskers, hillocks, irregular growths, popped grains, and sunken grains) shown in their corresponding TC histograms.

The defect densities reported in Table 7.6 for SonW and MonW are only for defects that have grown out of the plane. In addition, the total volumes of defects for thermal cycling Sn on W in Table 7.6 are the *net* volumes relaxed during thermal cycling. The net volume is obtained by subtracting the volume of the sunken grains from the total volume relaxed by whiskers, hillocks, irregular growths, and popped grains. Therefore, the net defect volumes for the three plating chemistries range from 11,423 to 12,807 μm^3 and represent the net compressive strain relieved by defects above the free surface. This narrow range of net volume suggests that each sample relieves approximately the same net compressive strain. There are, however, major differences in the defect volumes for each film type, as seen in Table 7.7, which excludes sunken grains. For SonW and MonW, the volumetric strains relaxed by growth out of the plane of the film are almost an order of magnitude higher than the net strains, in contrast to BonW, which showed minimal grain sinking. More detailed microstructural analysis is needed to understand the defect and grain shrinkage evolution.

TABLE 7.7 Summary of Defect Volumes, Film Thicknesses, and Strains that Exclude Sunken Grains for Sn-on-Tungsten Substrates Sorted by Film Type Bright (BonB), Satin (SonB), and Matte (MonB) for Thermally Cycled Samples

Surface Finish	Excluding Sunken Grains		
	Defect Volume ($\mu m^3/mm^2$)	Film Thickness (μm)	Strain (mm^3/mm^3)
BonW	11,569	5	2.31E−03
SonW	68,543	3.2	2.14E−02
MonW	97,049	3.6	2.70E−02

7.5 CONCLUSIONS

The surface-defect counting technique presented here provides a method for quantifying heterogeneous stress relaxation in tin films by the formation of whiskers, hillocks, irregular growth, popped grains, and sunken grains. The measurements are based on characterization of individual defects by defect morphology type and defect dimensions. From these measurements, the defect volume relaxed by defect formation and corresponding volumetric strain are calculated for a normalized 1 mm^2 area. By comparing data sets with systematic changes in films, for example, the changes in a specific film type as a function of stress state, changes in the active sites for stress relaxation can be followed.

The utility of this technique in quantifying differences between films was demonstrated using a suite of samples provided by the iNEMI Tin Whisker Assessment Project. Eighteen sample types were prepared using three types of electroplated tin films on two different substrates (brass and tungsten) and exposed to three different environments: 2-year ambient storage, thermal cycling for 500 cycles from −45 to 85°C, and temperature/relative humidity for 1000 h at 50°C/50%RH. Initial analyses of the data sets demonstrate the complex interplay between the film/substrate stress relaxation mechanisms, film microstructure, and stress conditions.

APPENDIX

Figures 7.A1–7.A10.

ACKNOWLEDGMENTS

The authors are grateful for the collaboration with the iNEMI Tin Whisker Assessment project team members Earl Miller, Richard Parker, Peng Su, and Tom Woodrow, for advice and guidance from Bill Russell of Texas Instruments on statistics, for funding from SAIC and Crane/NSWC for this project, and for the contributions of Purdue University – Merrillville interns Michael Weber and Antonio Palmerin.

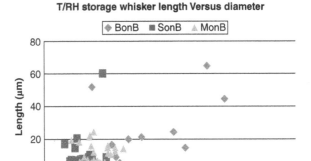

Figure 7.A1   Scatter plot of whisker growth on bright, satin, and matte Sn-electroplated films on brass substrates after T/RH for 1000 h at 50°C/50%RH.

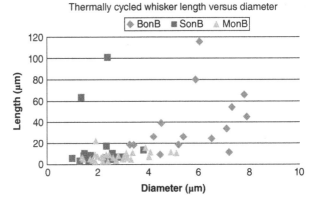

Figure 7.A2   Scatter plot of whisker growth on bright, satin, and matte Sn-electroplated films on brass substrates after thermal cycling for 500 cycles from −45 to 85°C.

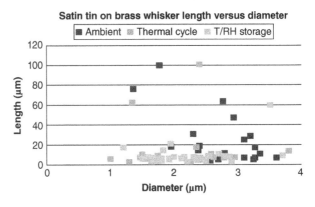

Figure 7.A3   Scatter plot of whisker growth on satin Sn-electroplated films on brass substrates at three different storage conditions: 2 years at ambient, TC = 500 cycles from −45 to 85°C, T/RH = 1000 h at 50°C/50%RH.

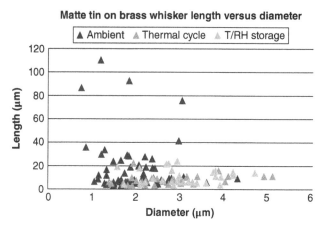

Figure 7.A4 Scatter plot of whisker growth on matte Sn-electroplated films on brass substrates at three different storage conditions: 2 years at ambient, TC = 500 cycles from −45 to 85°C, T/RH = 1000 h at 50 °C/50%RH.

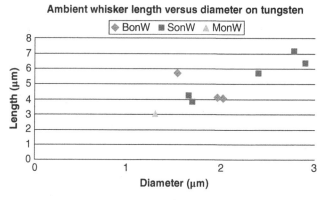

Figure 7.A5 Scatter plot of whisker growth on bright, satin, and matte Sn-electroplated films on tungsten substrates after 2 years of ambient storage.

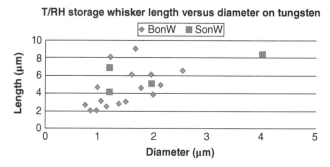

Figure 7.A6 Scatter plot of whisker growth on bright and satin Sn-electroplated films on tungsten substrates after T/RH for 1000 h at 50°C/50%RH.

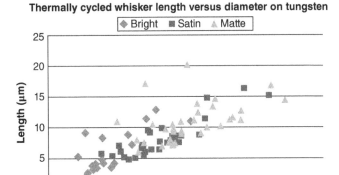

Figure 7.A7 Scatter plot of whisker growth on bright, satin, and matte Sn-electroplated films on tungsten substrates after thermal cycling for 500 cycles from −45 to 85°C.

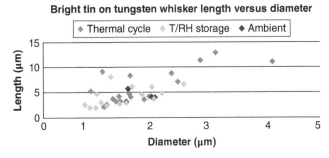

Figure 7.A8 Scatter plot of whisker growth on bright Sn-electroplated films on tungsten substrates at three different storage conditions: 2 years at ambient, TC = 500 cycles from −45 to 85°C, T/RH = 1000 h at 50°C/50%RH.

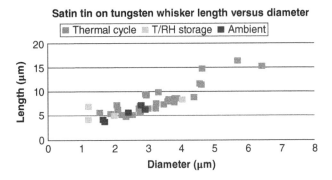

Figure 7.A9 Scatter plot of whisker growth on satin Sn-electroplated films on tungsten substrates at three different storage conditions: 2 years at ambient, TC = 500 cycles from −45 to 85°C, T/RH = 1000 h at 50°C/50%RH.

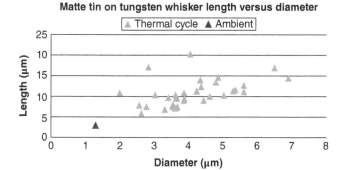

Figure 7.A10 Scatter plot of whisker growth on matte Sn-electroplated films on tungsten substrates at two different storage conditions: 2 years at ambient, TC = 500 cycles from −45 to 85°C.

REFERENCES

1. JEDEC standard JESD22A121. *Measuring Whisker Growth on Tin and Tin Alloy Surface Finishes.* JEDEC; 2008.

2. JEDEC standard JESD201. *Environmental Acceptance Requirements for Tin Whisker Susceptibility of Tin and Tin Alloy Surface Finishes.* JEDEC; 2006.

3. Lee CJ, Reynolds H. Tin whisker accelerated test project. Electronic Materials and Packaging, 2006. EMAP 2006. International Conference, 2006 December; IEEE. pp. 1–21, 11–14.

4. Osenbach JW, Reynolds HL, Henshall G, Parker RD, Su P. Tin whisker test development—temperature and humidity effects. Part II: Acceleration model development. IEEE Trans Electron Packag Manuf 2010;33(1):16–24.

5. Reynolds HL, Lee CJ, Smetana J. In: Bradley E, Handwerker C, Bath J, Parker R, Gedney R, editors. *Tin whiskers: mitigation strategies and testing, Chapter 7, Lead-Free Electronics.* IEEE; 2007.

6. Reynolds HL, Osenbach JW, Henshall G, Parker RD, Su P. Tin whisker test development—temperature and humidity effects. Part I: Experimental design, observations, and data collection. IEEE Trans Electron Packag Manuf Jan. 2010;33(1):1–15.

7. Schroeder V, Bush P, Williams M, Vo N, Reynolds HL. Tin whisker test method development. IEEE Trans Electron Packag Manuf Oct. 2006;29(4):231–238.

8. Panashchenko L. Evaluation of environmental tests for tin whisker assessment. Masters Degree, Mechanical Engineering, University of Maryland (College Park, MD), Digital Repository at the University of Maryland, 2009.

9. Rasband WS. ImageJ, U. S. National Institutes of Health, Bethesda, Maryland, USA, http://imagej.nih.gov/ij/, 1997–2014.

10. Sarobol P, Wang Y, Chen WH, Pedigo AE, Koppes JP, Blendell JE, Handwerker CA. Chapter 1 in this book; 2015.

11. Chason, E, Jadhav, N. Chapter 2 in this book; 2015.

12. Clore NG. Identification and counting procedure for tin whiskers, hillocks, and other surface defects. Masters Degree, Mechanical/Materials Science and Engineering, Purdue University, 2011.

13. Chen WH, Holaday JR, Handwerker CA, Blendell JE. Effect of crystallographic texture, anisotropic elasticity, and thermal expansion on whisker formation in beta-Sn thin films. J Mater Res Jan 2014;29(2):197–206.

8

BOARD REFLOW PROCESSES AND THEIR EFFECT ON TIN WHISKER GROWTH

JASBIR BATH

Bath Consultancy LLC, San Ramon, California, USA

8.1 INTRODUCTION

This chapter reviews the potential affect that the board reflow process has on tin whisker growth. Investigations in this area have looked at grain orientations and grain size, solder paste volume, and solder peak temperature on tin whisker growth. The results on tin whisker testing on soldered pure-tin-coated components after reflow are discussed in the following sections.

8.2 THE EFFECT OF REFLOWED COMPONENTS ON TIN WHISKER GROWTH IN TERMS OF GRAIN SIZE AND GRAIN ORIENTATION DISTRIBUTION

To understand the effect of reflowed components on tin whisker growth, tin whisker tests have been done on as-received and reflowed 0.5-mm-pitch LQFP64 pure-tin-plated components [1]. For this component, there was little difference between packages before and after SnAgCu reflow in terms of whisker length and whisker density after ATC testing from −55 to +85°C. There was a reduction in tin whisker length for the LQFP64 component after SnAgCu reflow after temperature

Mitigating Tin Whisker Risks: Theory and Practice, First Edition.
Edited by Takahiko Kato, Carol A. Handwerker, and Jasbir Bath.
© 2016 John Wiley & Sons, Inc. Published 2016 by John Wiley & Sons, Inc.
Companion website: www.wiley.com/go/Kato/TinWhiskerRisks

and humidity testing (60°C/93%RH for 3000 h) from 40 μm for as-received down to 5 μm for after reflowed components.

Tin whisker tests were also done on as-received and reflowed 0.65-mm-pitch PQFP132 pure-tin-plated components. For the PQFP132 component, there was a reduction in whisker length and whisker density on packages after SnAgCu reflow after ATC testing from −55 to +85°C from 25 μm for as-received components to 15 μm after reflow. Whisker density reduced from 100 whiskers per lead for as-received components to less than 30 whiskers per lead for reflowed components after 3000 ATC cycles.

There was also found to be a reduction in tin whisker length for these components after SnAgCu reflow after temperature and humidity testing (60°C/93%RH for 3000 h) from 70 μm for as-received to 10 μm for reflowed components.

The SnAgCu reflow process changed the overall microstructure of the coated component with the grain size increased after reflow with the whisker growth reduced after reflow based on this. Grain orientation distribution before SnAgCu reflow was also different than after SnAgCu reflow for the PQFP132 component, which had an affect on whisker growth.

Reflow process changes affected the grain orientation distribution and grain size of the tin finish, which could affect the stress distribution in the coating and thus whisker growth propensity. The tin whisker length after SnAgCu reflow after ATC testing was similar for both LQFP64 and PQFP132 component packages (10–15 μm). The tin whisker length after SnAgCu reflow after temperature and humidity testing was similar for both component packages (5–10 μm).

8.3   REFLOW PROFILES AND THE EFFECT ON TIN WHISKER GROWTH

8.3.1   Reflow Standards Related to Component Tin Whisker Growth Testing

For technology acceptance tin whisker testing of components, certain component test samples have to be preconditioned with a tin–lead and a lead-free reflow profile before tin whisker testing based on JEDEC JESD201 standard [2]. The actual reflow profile to be used is referenced in JEDEC 22-A121 standard [3]. For both tin–lead and lead-free reflow profiles, the temperature is measured on the component lead or solder joint or at the surface of the test coupon, which is in contact with the component. This temperature measurement location is different, compared with that used in the component moisture sensitivity temperature rating standard J-STD-020 [4], which measures the component package body temperature.

For the tin–lead reflow profile used in JEDEC 22-A121 standard [3], the preheat temperature is 60–120 s from 100 to 150°C with a time over 183°C of 60–120 s. The peak component lead or solder joint temperature should be between 200 and 220°C, which is meant to ensure that the pure tin component coating with a melting point of 232°C is not melted during the preconditioning temperature profile.

For the lead-free reflow profile used in JEDEC 22-A121 standard[3], the preheat temperature is 60–120 s from 150 to 200°C with a time over 217°C of 60–120 s. The peak component lead or solder joint temperature should be between 245 and 260°C, which is meant to ensure that the pure tin component coating with a melting point of 232°C is melted during the preconditioning.

JEDEC JESD201 standard [2] also indicates that for technology acceptance tin whisker testing component preconditioning, should use a peak temperature for both tin–lead and lead-free reflow at the lower end of the range indicated to avoid substantial wetting of the terminations by the reflowed solder paste to promote unwetted areas of the component termination where tin whisker growth is more likely to occur. It indicates to use a low-activity flux-type solder paste in air reflow atmosphere and to apply less solder paste on the board using reduced stencil aperture openings and/or reduced stencil thickness. It also recommends to clean the flux residues on the assembled test boards prior to tin whisker testing due to the unknown affect of reflowed flux residue from solder paste on tin whisker growth.

8.3.2 Reflow Profile Standards Related to Electronics Board Assembly

In addition to the reflow profile standards related to the component lead frames for tin whisker preconditioning, component temperature standards in general should be reviewed. J-STD-020 standard [4] discusses the component package temperature rating of nonhermetically sealed moisture-sensitive components.

For tin–lead component package temperature ratings used in J-STD-020 standard [4] based on package thickness and volume, they vary from testing at three times 220°C peak to three times 235°C peak temperature. In terms of the time above 183°C, this is between 60 and 150 s at the component package, which is different from JEDEC 22-A121 standard [3], which indicates the time above 183°C of between 60 and 120 s at the component lead or solder joint or at the surface of the test coupon, which is in contact with the component lead. J-STD-020 standard [4] also indicates the time within 5°C of the component package peak temperature of 20 s for tin–lead component packages.

For lead-free component package temperature ratings used in J-STD-020 standard [4] based on package thickness and volume, they vary from testing at three times 245°C peak to three times 260°C peak temperature. An additional 260°C profile is tested for area array components to simulate lead-free rework for BGA/CSP parts components, which are not rated to 260°C peak temperature. In terms of the time above 217°C, this is between 60 and 150 s at the component package, which is different from JEDEC 22-A121 standard [3], which indicates the time above 217°C of between 60 and 120 s at the component lead or solder joint or at the surface of the test coupon, which is in contact with the component lead. J-STD-020 standard [4] also indicates the time within 5°C of the component package peak temperature of 30 s for lead-free component packages.

J-STD-075 standard [5] refers to the Assembly Classification for Non-IC Components such as Chip components, Aluminum capacitors, Crystals, Oscillators, Fuses, LEDs, and Relays. It classifies reflowed components based on a R0 to R9 identification. R0 indicates that the component is not process sensitive, whereas R4 classifies the component at three times 260°C peak temperature and R7 classifies the component at three times 245°C peak temperature. Additional labeling also indicates Reflow Time limitations and other process-related issues.

There are some differences between temperatures at the solder joint, component lead frame, and component packages. Usually, the temperature differences are minimal between the solder joint and component lead frame with a greater potential difference between these two locations and the component package body especially if the package body size is large.

8.3.3   Tin–Lead(SnPb) and Lead-Free(SnAgCu) Reflow Profiles

The major difference between the tin–lead and lead-free reflow profile is the different preheat temperatures and time and peak temperatures and times. These can have an effect on tin whisker growth as mentioned in the previous sections.

In tin–lead soldering assembly, the typical soak/preheat stage is between 120 and 183°C from 60 to 180 s with a peak temperature of 205–215°C with time over 183°C of 45–90 s. These are lower in peak temperature and time than tin–lead component temperature ratings referred to in J-STD-020 standard [4] as the component temperatures and times are the maximum that the component can withstand based on the component supplier testing and the fact that there should be some margin between the typical processing temperatures during production and the absolute maximum temperature ratings of the components.

In lead-free Sn3Ag0.5Cu soldering assembly, the typical soak/preheat stage is between 150 and 217°C from 60 to 180 s with a peak temperature of 235–255°C with time over 217°C of 45–90 s. Again, these are lower in peak temperature and time than lead-free component temperature ratings referred to in J-STD-020 standard [4] for the same reasons as described for tin–lead soldering assembly.

8.3.4   Optimizing the Assembly Reflow Profile Process

Optimizing the reflow process usually depends on the temperature requirements for the solder paste, the components to be soldered, and the board temperature rating. The profile is developed to take into account these factors as well as to ensure a low-temperature delta across the board and produce good soldering results.

In terms of the solder paste, generally a higher soldering peak temperature and time above the melting point is preferred to ensure good soldering results, but too high a temperature can cause solder joint voiding, increased intermetallic compound growth in the solder joint potentially affecting solder joint reliability, and component and board temperature issues. The component and board temperatures should be kept as low as possible to prevent high-temperature issues while still ensuring good soldering results.

In terms of technology qualification of components before tin whisker testing, a lower soldering peak temperature is preferred to have unwetted areas on the component lead frame where tin whiskers may occur. As lead-frame components have more finer pitches, the amount of solder volume at the solder joint would be less, which may also produce more unwetted areas, increasing the tendency for tin whisker growth.

8.4 INFLUENCE OF REFLOW ATMOSPHERE AND FLUX ON TIN WHISKER GROWTH

Work conducted by Baated et al. [6] showed the affect of reflow atmosphere and flux activity on tin whisker growth. Evaluations were conducted with pure-tin-coated Cu and Alloy 42 QFP lead-frame components soldered with Sn3Ag0.5Cu solder paste. Solder pastes used were halogen-free and those containing 0.4 wt% (4000 ppm) Br and 0.8 wt% (8000 ppm) Br. Reflow atmosphere was air or nitrogen (<500 ppm O_2) with a reflow peak temperature of 240°C.

Assembled test board components were then subjected to 1000 h at 85°C/85%RH followed by tin whisker inspection on the soldered components. Tin whiskers were observed with 0.4 wt% Br and 0.8 wt% Br solder paste reflowed joints in air reflow atmosphere. The tin whisker length was <5 μm with 0.4 wt% Br solder paste for both Cu and Alloy 42 lead-frame QFP components. The tin whisker length was <10 μm with 0.8 wt% Br solder paste on Cu lead-frame components and <15 μm with 0.8 wt%Br solder paste on Alloy 42 lead-frame components. The tin whisker growth was related to the increased oxidation of the soldered joint based on increased solder flux paste activity and air reflow atmosphere.

Tsukui et al. [7] investigated the effect of different flux activators on tin whisker growth on Sn3Ag0.5Cu soldered pads. The flux activators tested included diethyl amine HBr salt (2 wt% and 4 wt% concentrations), 1.9 wt% adipic acid, and 2.41 wt% lauryl amine. The soldered pads were fluxed with the different activators, then reflowed at 230°C peak temperature, followed by testing at 85°C and 85%RH.

After 2000 h of testing, 2 wt% HBr showed a maximum tin whisker length of 83 μm versus 87 μm with 4 wt% HBr. After 10,000 h of testing, 2 wt% HBr showed tin whiskers with a maximum length of 208 μm. After 7000 h of testing, 1.9 wt% adipic acid showed a maximum tin whisker length of 24 μm versus 70 μm with 2.41 wt% lauryl amine. For the Sn3Ag0.5Cu soldered pads with no flux activator, the maximum tin whisker length was 9 μm after 10,000 h of testing at 85°C and 85%RH.

The oxidation of the lead-free Sn3Ag0.5Cu solder was aided by the different flux activators, which increased tin whisker growth after temperature and humidity testing.

Work conducted by Snugovsky et al. [8] showed the affect of chloride (Cl^-) and sulfate (SO_4^{2-}) contamination on assembled SOT23, QFP44, SOIC8, SOIC20, and PLCC44 components with Sn3Ag0.5Cu solder. The Cl contamination was 100 ppm, and the SO4 contamination was 243 ppm.

SOT23 Alloy 42 lead-frame pure-tin-coated components had tin whiskers of 20 μm long with 100 ppm Cl and 23 μm long with 243 ppm SO4 after 500 h at 85 °C and 85%RH. 0.8-mm-pitch QFP44 copper alloy (C7095) lead-frame pure-tin-coated components had tin whiskers of 39 μm long with 100 ppm Cl and 40 μm long with 243 ppm SO4. The QFP44 components had tin whiskers of 6 μm when there was no contamination with Cl or SO4.

SOIC8 copper alloy (C194) lead-frame pure-tin-coated components had tin whiskers of 22 μm long with 100 ppm Cl and 243 ppm SO4 after 500 h at 85°C and 85%RH. SOIC20 copper alloy (C194) lead-frame pure-tin-coated components had tin whiskers of 45 μm long with 100 ppm Cl and 243 ppm SO_4 after 500 h at 85°C and 85%RH.

PLCC44 copper alloy (C151) lead-frame pure-tin-coated components had a tin whisker of 160 μm long with 100 ppm Cl and 243 ppm SO_4 after 500 h at 85°C and 85%RH.

Increased contamination of the soldered components increased tin whisker growth after the temperature and humidity testing.

8.5 EFFECT OF SOLDER PASTE VOLUME ON COMPONENT TIN WHISKER GROWTH DURING ELECTRONICS ASSEMBLY

In addition to the solder paste reflow profile, the amount of solder paste printed at the component board pad locations has an effect on tin whisker growth. A study using SOT3 component A and SOT3 component B, which both had the same Alloy 42 lead frames but different matte Sn component finish from different component suppliers, was done to evaluate this effect [9].

Stencil aperture and stencil thickness were varied to adjust the amount of solder paste volume printed on the board pads. The solder paste amount printed was adjusted to 1:1 with board pad (100%), 30% less, 20% less, 10% less, 10% greater, and 20% greater.

Both loose components and components assembled with either tin–lead or Sn3Ag0.5Cu paste were thermally cycled from −55 to +85°C for 1500 cycles with 20 min duration for each thermal cycle. The Sn37Pb and Sn3Ag0.5Cu solder pastes used were no-clean with low flux activity (ROL0). Reflow of the assembled components with the solder pastes was done in a nitrogen atmosphere.

Peak temperatures used for tin–lead solder paste soldering were 205, 215, and 225°C. Peak temperatures used for lead-free SnAgCu solder paste soldering were 235, 245, and 255°C. For the assembled components, there were 18 peak temperature/paste volume combinations each for SOT3 components A and B with both the Sn37Pb and Sn3Ag0.5Cu solder pastes. There were six components tested per test cell.

For loose unassembled components, the tin whisker length was up to 80 μm for SOT3 component A after 1500 ATC cycles versus up to 50 μm for SOT3 component B. The component fail limit for Class 2 tin whisker testing using JESD201 standard [2] was above 45 μm tin whisker length; therefore, both SOT3 components would have failed the test. Soldered paste assembled components generally had less whisker growth than loose components after ATC testing.

For Sn37Pb paste soldered components, the assembled SOT3 component B had less tin whiskers than assembled SOT3 component A. No tin whiskers were observed for component B at the higher 215 and 225°C peak temperatures and for component A at 225°C peak temperature with tin–lead solder paste. For component A at 215°C, whisker growth was only observed when the solder paste volume applied was 30% less than the board pad with 1:1 stencil to board pad paste volume.

When the peak temperature was lower for component B at 205°C peak, one test cell out of six (with pasted volume applied being 20% less than the board pad) showed tin whiskers that were less than 45 μm in length.

For component A at 205°C peak, all test cells had tin whiskers with four out of six test cells having tin whisker length over 45 μm with the other two of the six test cells less than 45 μm in length generally showing the trend of low paste volume having more tin whiskers with greater whisker lengths. Whisker growth was observed at the lead near the top of the package where the reflowed SnPb paste did not wet up. The vertical section of the lead where the SnPb solder wetted did not show whisker growth.

In general, the higher the tin–lead peak temperature, the less whisker growth due to the increased wetting of the SnPb solder paste at the package lead frame. Lower SnPb paste volume with a lower peak temperature produced more tin whisker growth because of less spreading of the reduced amount of solder paste applied.

For SnAgCu paste soldered components, assembled SOT3 component B generally had less tin whiskers than SOT3 component A. No tin whiskers were observed for SOT3 component B at 235 and 255°C peak temperature. At 245°C peak for assembled component B, there was evidence of some tin whiskers but they were less than 45 μm in length.

For SnAgCu assembled SOT3 component A, tin whiskers were observed at 235, 245, and 255°C peak temperatures with whiskers all less than 45 μm at 235 and 245°C peak.

At 255°C for SOT3 component A, there was one test cell that had tin whisker greater than 45 μm (with solder paste volume applied being 20% less than the board pad). There was no clear trend for whisker growth in terms of paste volume and peak temperature for SnAgCu soldered components as the wetting of the SnAgCu soldered components was similar at all three peak temperatures (235, 245, 255°C). If there was low paste volume as was the case at 255°C peak with paste volume applied 20% less than the board pad, this would help to increase tin whisker growth potential.

Whisker growth in all cases with the SnAgCu assembled components was at lead locations close to the body of the package as was the case with the SnPb assembled components. The wetting areas of the SnAgCu paste on the lead were similar to the SnPb paste with no wetting at the lead near the top of the package where the reflowed SnAgCu paste did not wet up. The vertical section of the lead where the SnAgCu solder wetted did not show whisker growth.

With increased solder paste wetting during reflow, the component lead-frame would be covered more with solder, which generally reduced the tendency for tin whisker growth.

8.6   CONCLUSIONS

There has been work to understand the affect of reflow profile, reflow atmosphere, solder paste volume, and flux activity on tin whisker growth.

In terms of solder paste volume, increased solder volume printed has been found to reduce the amount of unwetted areas of the component lead frame reducing tin whisker growth, which has been found to be especially the case for tin–lead versus lead-free solder paste.

For tin–lead soldering, the lower soldering peak temperature reduced solder paste wetting up the component lead frame increasing the potential for tin whisker growth.

For lead-free soldering, no-clear trend was seen between the soldering peak temperature and the potential for tin whisker growth as wetting was similar for different peak soldering temperatures from 235 to 255°C.

Air reflow atmosphere was found to increase the tendency for tin whisker growth in combination with increased lead-free solder paste flux activity based on increased oxidation of the soldered component joint.

Different flux activators were found to aid the oxidation of tin, which increased tin whisker growth after temperature and humidity testing. Increased contamination of the soldered components with chloride and sulfate also increased tin whisker growth after temperature and humidity testing.

ACKNOWLEDGMENTS

The author would like to acknowledge the persons and companies who contributed information from various papers and standards that were reviewed in this chapter.

REFERENCES

1. Su P, Ding M, Chopin S. Effects of reflow on the microstructure and whisker growth propensity of Sn finish. Proceedings of the 55th ECTC Conference; IEEE; 2005, pp. 434–440.
2. JEDEC JESD201 standard. Environmental Acceptance Requirements for Tin Whisker Susceptibility of Tin and Tin Alloy Surface Finishes, JEDEC.
3. JEDEC 22-A121 standard. Test Method for Measuring Whisker Growth on Tin and Tin Alloy Surface Finishes, JEDEC.
4. IPC/JEDEC J-STD-020 standard. Moisture/Reflow Sensitivity Classification for Nonhermetic Solid State Surface Mount Devices, IPC.
5. ECA/IPC/JEDEC J-STD-075 standard. Classification of Non-IC Electronic Components for Assembly Processes, IPC.
6. Baated A, Kim K-S, Suganama K, Huang S, Jurcik B, Nozawa S, Stone B, Ueshima M. Effects of reflow atmosphere and flux on tin whisker growth of Sn–Ag–Cu solder. SMTAI Conference; SMTA; 2009.
7. Tsukui T, Takeuchi Y, Ueshima M, Takenaka J, Takeuchi M, Kamiyama A, Sasaki K. Study on flux and alloy of lead-free solder with mitigation effect and consideration for acceleration test method. J Jpn Inst Electron Packag 2012;15(5):404–416.

8. Snugovsky P, Meschter S, Kapadia P, Romansky M, Kennedy J, Kosiba E. Influence of board and component cleanliness on whisker formation. IPC/SMTA High Performance Cleaning and Coating Conference; SMTA; 2010.

9. Su P, Lee C, Li L, Xue J, Khan B, Moazeni R, Hartranft M. Practical assessment of tin whisker growth risk due to environmental temperature variations. Proceedings of 59th ECTC; IEEE; 2009. pp 736–741.

9

MECHANICALLY INDUCED TIN WHISKERS

TADAHIRO SHIBUTANI

Yokohama National University, Yokohama, Kanagawa, Japan

MICHAEL OSTERMAN

Center of Advanced Life Cycle Engineering (CALCE), University of Maryland, College Park, Maryland, USA

9.1 INTRODUCTION

As the interconnect density increases and the circuit current decreases (as low as microamperes) in electronic systems, the reliability of electronic connectors has become more critical [1]. In the past several decades, major progress in the study of electrical contacts and connectors has been obtained. For example, mixed flowing gas laboratory systems have been developed for studying environmental corrosion, a wide range of surface analysis approaches have become available, fretting corrosion of electrical contacts has been better understood, and new connector designs and lead-free finishes have been developed.

A contact finish is used to provide protection for the base metal of a connector against corrosion, diffusion, and contact wear and to help establish and maintain a stable resistance at the contact interface. There are two kinds of contact finish materials for electrical connectors: noble metals and alloys and nonnoble metals and alloys. Noble metals and alloys include gold, gold alloys, palladium, and palladium alloys. Noble metals and alloys are the preferred coating materials for separable connectors in the electronics industry because of their low contact resistance and

Mitigating Tin Whisker Risks: Theory and Practice, First Edition.
Edited by Takahiko Kato, Carol A. Handwerker, and Jasbir Bath.
© 2016 John Wiley & Sons, Inc. Published 2016 by John Wiley & Sons, Inc.
Companion website: www.wiley.com/go/Kato/TinWhiskerRisks

corrosion resistance. Nickel, nickel alloys, silver, tin, and tin alloys are the most common nonnoble contact finishes. They exhibit a surface film due to natural aging mechanisms. The surface films are primarily responsible for the failure of electrical contacts in the field.

Tin–lead solder alloys have been the most commonly used nonnoble contact finish materials in the electronic industry due to their low cost and ease of manufacture. With appropriate design considerations, tin–lead alloys may often be successfully utilized as cost-effective alternatives to gold under lower requirements of reliability. Since the tin–lead alloys are soft, the brittle oxides of tin–lead alloys can be easily displaced by proper mechanical deformation and the wiping action of contact surfaces, low contact resistance can be obtained. Extensive studies have been made on the electrical performance of tin–lead solder coatings. Tin and tin alloys are soft metals on which a thin but hard oxide layer is rapidly formed. Supported by a soft substrate, this layer is easily broken and its fragments can be pressed into the underlying matrix of soft, ductile tin. Over time, the embedded oxide can result in increased contact resistance and failure.

Driven by legislative requirements and market forces, lead-free solders are replacing tin–lead in soldering processes with lead-free solder alloys also being used in electronic connectors. The electronics industry has widely adopted pure (matte) tin and high-tin alloy finishes as a lead-free option. However, both silver and copper, which are present in popular lead-free alloys (Sn–Ag–Cu, Sn–Ag, Sn–Cu), are subject to film formation and corrosion in atmosphere, especially with corrosive pollutants such as H_2S and SO_2. The high tin content (>96%) in the alloys may exhibit poor fretting corrosion behavior. In addition, the adoption of high-tin content finishes has also created a reliability concern pertaining to the formation of conductive tin whiskers, which can bridge adjacent conductors and lead to current leakage and electrical shorts. Since the spacing between conductors is of the order of a few hundred microns, and since tin whiskers have been known to grow up to a few millimeters in length [2], tin whiskers pose a serious risk to connector reliability.

External pressure can induce tin whiskers and accelerate their growth [3, 4]. In the case of fine-pitch connectors with pure tin or high-tin alloy finishes, mating pressure between the connector elements can produce pressure-induced tin whiskers. Figure 9.1 shows a picture of tin whiskers induced by contact forces. A lot of whiskers are found around the contact area. Therefore, pressure-induced whiskers are a major concern in separable components. Fisher et al. [3] concluded that the growth rate is proportional to the pressure based on diffusion theory. Since stresses in the plating increase with pressure, the atom transport forming tin whiskers is accelerated by the gradient of stress.

Recent industrial studies show that the nature of pressure-induced whiskers differs from that of spontaneous whiskers as mentioned earlier [5, 6]. Filamentary or columnar whiskers are usually found near the pressured area. For spontaneous whiskers, temperature and humidity often accelerate the growth [7, 8]. On the other hand, pressure-induced whiskers grow rapidly at room temperature and no accelerated test has been developed. Furthermore, IMC growth is not usually found or necessary along grain boundaries to produce stresses in the plating because the whisker growth is

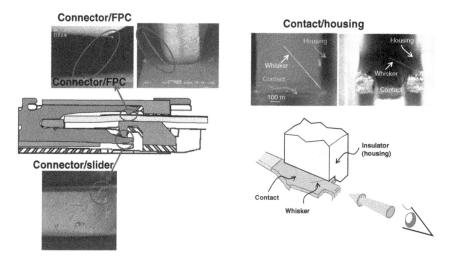

Figure 9.1   Pressure-induced tin whisker on a connector. Please visit www.wiley.com/go/ Kato/TinWhiskerRisks to access the color version of this figure.

too rapid. These facts suggest that there is another mechanism for pressure-induced whiskers.

In this chapter, the pressure-induced whisker at separable interfaces is discussed as a key reliability issue of lead-free separable contacts and connectors. Theories of mechanically induced tin whiskers are shown. Then, several case studies with pure tin and lead-free finishes were performed to evaluate pressure-induced tin whiskers.

9.2   OVERVIEW OF MECHANICALLY INDUCED TIN WHISKER FORMATION

Fisher et al. [3] were the first to report enhanced growth rate of tin whiskers under the pressure applied on tin-finished surfaces. In tests with compressive forces up to 52 MPa, whisker growth rates were documented to increase to as high as 10^4 A/s. Higher growth rates under higher pressures exhibited shorter growth periods than lower growth rates observed under lower pressures. They estimated a constant growth rate proportional to the applied pressure with an abrupt transition. Further, they ruled out extrusion due to the fact that single-crystal growth was observed. Later, Pitt and Henning [9] found that the growth rate was not constant but decayed with increasing time. They also disputed the maximum growth rate finding a maximum initial growth rate of 593 A/s at 55 MPa with an average rate of 195 A/s. They concluded that whisker growth was driven by compressive stress.

Yang and Li [10] found a correlation of whisker growth with load and plating thickness. Tests were conducted by application of loads between 50 and 400 g with a stainless steel 1.5 mm spherical tip to tin plate thickness over copper, which varied between 1 and 6 μm. In this study, increasing the plating thickness decreased whisker growth. Increased indention load increased the frequency and length of whiskers. Measurements taken between 1 and 15 days demonstrated an increase in whisker frequency and length with time.

Lin and Lin [11] conducted whisker growth by applying a compressive stress using a four-point bend fixture. Estimated outer surface in-plane stress ranged between −15.5 and 7.75 MPa. Tests indicated an optimum compressive stress for producing long whiskers with higher compressive stress nucleating more whiskers rather than longer whiskers. In this study, the application of a tensile stress in the film was shown to reduce whisker growth, compared to whisker growth on nonstressed samples. Chen and Chen [12] also found reduced whisker formation on tin samples placed under 22 MPa tensile stress, compared to similarly plated nonstressed samples.

Southworth et al. [13] demonstrated an optimum strain level for whisker growth. In this study, 7% compressive strain in the tin film was found to produce the longest whisker with 1.5% and 16% compressive strains found to produce shorter whiskers.

Moriuchi et al. [14] defined an indention test method and provided results for application of 100 and 200 g loads using a 1 mm spherical indenter. In this study, a 3-μm-thick tin–copper plating over 2-μm -thick nickel underlayer on phosphor bronze was tested. Whisker growth increased with time and load. The longest reported test time of 120 h was found to have the longest whiskers with the highest 200 g load.

9.3 THEORY

9.3.1 Creep-Based Model

A creep-based tin whisker model was proposed to explain experimental data [4]. Figure 9.2 shows the creep-based tin whisker model. When a contact load is applied

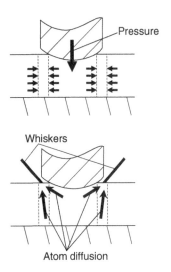

Figure 9.2 Schematic of creep-based tin whisker model.

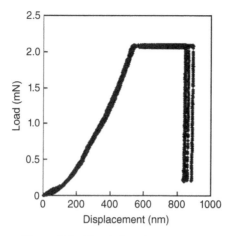

Figure 9.3   Nanoindentation creep test.

to a surface of the plating, the material below the pressured area is expanded in the orthogonal directions (Poisson effect) due to the deformation of the plating. Since the melting point of tin plating is relatively low, the plating creeps even at room temperature. This process induces a multiaxial stress concentration outside the pressured area. Although stress relaxation occurs, multiaxial stress components remain as residual stresses, which act on grain boundaries of the plating. These residual stresses can cause stress-induced atom migration and the formation of whiskers. The model has been verified on the basis of reported industrial studies.

To assess the plating for pressure-induced whiskers, a creep property of the plating has to be measured. A nanoindentation technique can be used to measure the creep property in the plating. The indentation process consists of loading, holding, and unloading (see Fig. 9.3). During the holding segment, the increasing indentation displacements due to creep are recorded. The resulting stress is proportional to hardness, and the strain rate is the indenter descent rate divided by the current contact depth. Thus, the relationship between indented displacement u and hardness H as a function of time can be expressed by [15–17]

$$\frac{\dot{u}}{u} = A(\alpha H)^n \qquad (9.1)$$

where A and n are the creep constant and exponent and α is a material constant. In this study, α is 1/3. The hardness, H, comes from the applied load and measured displacements. Thus, A and n can be obtained by fitting the data to the $\dot{u}/u - H$ curve. In this study, the strain rate $\dot{u}/u$ usually varies from 10^{-2} to 10^{-6}.

Seven tin and tin-alloy plated connectors listed in Table 9.1 were assessed. All finishes were plated on a nickel underlayer to prevent diffusion of copper from the substrate. These industrial samples have different thickness. The thickness of each finish ranged from 1 to 2 μm. All tests were conducted at room temperature by using

TABLE 9.1 Connector Finishes Subject to Nanoindentation Tests

ID	Finish on Connectors	Observed Whisker Length
A	Bright Sn–10Pb	Less than 50 μm
B	Semibright Sn–1.5Cu	Over 200 μm
C	Semibright Sn	
D	Matte Sn	
E	Reflowed Sn	Ranges from 50 to 150 μm
F	Matte Sn–2Bi	
G	Matte Sn–2Ag	

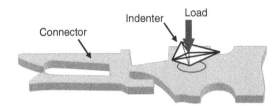

Figure 9.4   Nanoindentation test for the plating on the connector.

a MTS Nano-indenter XP. A Berkovich indenter was used. The angle of three edges on the pyramidal tip to the surface of the sample is 12.95°. Figure 9.4 shows the schematic of tested samples and indented position. Indentation was performed on the side of connectors as shown in Figure 9.4. To avoid the effect of the substrate, the initial indentation depth at the beginning of the holding segment was set to within 20% of the plating thickness. For each sample, at least five tests were carried out and the creep exponent was taken as the average value.

Figure 9.5 shows the creep exponent of each finish. The creep exponent characterizes the deformation mechanism of the plating. The value of the creep exponent ranges widely from 6 to 22, although the value of the bulk metal ranges only from 4 to 6. This result implies that the different samples have different deformation mechanisms.

Tin–copper plating has the highest value of creep exponent. For a higher creep exponent, stress relaxation will occur quickly, and higher residual stresses will remain. On the other hand, tin–lead alloy plating has the lowest creep exponent. This means lower stresses remain in the tin–lead plating. That is, the creep property affects the stress level in the plating.

The creep exponent of semibright pure tin (C) is higher than that of matte pure tin (D). Since there is a difference in the grain size between two finishes, the creep property can be also changed. Upon comparison between semibright finishes, the creep exponent of Sn–1.5Cu (B) is twice as high as that of pure tin (C). It suggests that material composition can affect the creep behavior even if grain size is controlled. Addition of Cu to pure tin plating is known to affect the microstructure of the plating

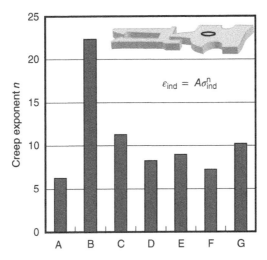

Figure 9.5   Creep exponent of various tin and tin–lead plating on connectors; A to G correspond to the samples listed in Table 9.1.

finish [18]. Cu-based precipitates along grain boundaries are considered to cause the
lack of grain boundary mobility.

Additions of Ag and Bi to matte pure tin induce slight changes in the value of
creep exponent. According to the JEITA report [6], there is no significant difference
in grain structure, but some precipitates are found along grain boundaries. Thus, material composition affects the creep properties of the plating finish.

Tin whisker tests for the above seven connectors were carried out by JEITA
[5]. Each connector was mated to a flat flexible cable. All samples were exposed
to various environmental conditions (Ambient, 15°C/50%RH, 25°C/50%RH,
35°C/50%RH, and 60°C/93%RH). After 1000 h storage, observation of tin whiskers
was conducted by using optical microscope and SEM. The longest filamentary
whiskers were found on the tin–copper plating as shown in Figure 9.6. In this
assessment, the length of whiskers was found to be less than 50 μm on the tin–lead
plating. On the other samples, the length of the filamentary or columnar whiskers
ranged from 50 to 150 μm. The creep exponents shown in Figure 9.5 correspond to
pressure-induced tin whisker formation. The data implies that the creep property
provides a screen for finishes. Comparing the pure tin finishes, semibright tin plating
(C) has a higher creep exponent than does matte tin (D) or reflowed tin (E).

9.3.2   Threshold Stress for Tin Whisker Growth

Stresses should overcome the increase in surface energy of a tin whisker to continue
the growth. Then, there is a threshold stress to produce tin whiskers. Since the surface
energy depends on the diameter of tin whiskers, the threshold stress may be also
related to the diameter of whiskers.

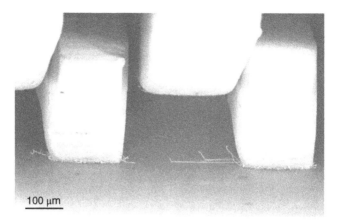

Figure 9.6   Tin whiskers growing from tin–copper finished connectors after 1000 h storage for the mating test.

Measurement of stresses is a key to qualify tin whisker formation. Lee and Lee measured the residual stresses of 7-μm-thick pure tin finish on a 70-μm-thick phosphor bronze coupon by using a deflection method. The value of compressive stress reached 8 MPa after 7 days [19]. Filamentary whiskers grew from the surface of the pure tin finish. Boettinger et al. [18] also measured initial residual stresses of Sn, Sn–2%Pb, and Sn–3%Cu by using the deflection method. The highest value of residual compressive stress was 36.5 MPa for 16-μm-thick Sn–3%Cu finish. Filamentary whiskers were found on the surface of Sn–3%Cu. On the other hand, compressive stress of 15.5 MPa was developed in pure tin finish, but only hillocks were found on the surface. Residual stress in tin or tin-alloy finish can be measured by using X-ray diffraction (XRD). It was reported that the bright tin finish (small grain size) on the Cu substrate has 10–15 MPa of compressive stress and that matte tin (large grain size) on Cu has 5–10 MPa.

This subsection presents a fundamental method to estimate the threshold stress to generate tin whiskers. A model to estimate a threshold stress of tin whisker formation is based on the balance of energy.

A simple whisker formation model in Figure 9.7a is used to explain the relation between stresses and whisker size. Whisker growth produces new surface area on the side of the whisker. Stresses to generate whiskers have to overcome the increase in surface energy. When a whisker of the size $2a$ grows up by the length of $\Delta \ell$, the increase in surface energy ΔU_s can be expressed as follows:

$$\Delta U_s = \gamma_s \cdot (4\sqrt{2}a) \cdot \Delta \ell \qquad (9.2)$$

γ_s is the surface free energy per unit area. When a pressure p is applied on the bottom of the whisker, external work consists of pile-up due to accumulated atoms and a loss of friction in sliding. Total work can be expressed as follows:

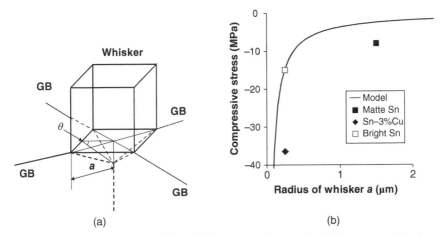

Figure 9.7 Relation between radius of whisker and estimated threshold compressive stress. (a) Tin whisker model. (b) Threshold stress for tin whisker growth.

$$W_p = \int_A (p \cdot \Delta\ell \cos\theta - \mu p \cdot \Delta\ell \sin\theta)dA \qquad (9.3)$$

where μ is the friction coefficient on the boundary and A indicates four grain boundaries at the bottom of the whisker. Then, the whisker can grow up on satisfying the following condition:

$$W_p > \Delta U_s \qquad (9.4)$$

If the pressure is assumed to be constant as the average value, the pressure has to satisfy the following equation:

$$p > \frac{2\sqrt{2}\gamma_s}{a(1 - \mu\tan\theta)} \qquad (9.5)$$

Figure 9.7b shows the typical relation between whisker radius and pressure. Here, γ_s of 0.55 J/m^2 is taken from Ref. [18]. μ is $1/\sqrt{3}$, which is the maximum value from the shear yield condition. θ is $\pi/4$ calculated from a balance of grain boundary energies between whisker and vertical grains. The critical stress to produce whisker growth is drastically changing in less than 0.5 μm. It implies that high compressive stresses are required in a finish of small grains such as the bright finish. Several experimental data are also plotted in this chart. Radii of whiskers are taken from Refs [18, 19] or typical values (bright tin: 0.25 μm, matte tin: 1.5 μm).

Nanoindentation tests were conducted at room temperature by using the MTS Nano IndenterXP. A three-sided Berkovich tip was used. The angle of three edges on the pyramidal tip to the surface of the samples is 12.95°. The indentation process consists of loading, holding, and unloading processes. During the loading process, applied load is up to the maximum load in 15 s. The values of applied load

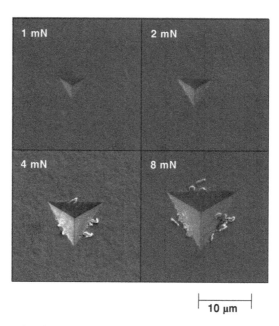

Figure 9.8 Indented surfaces of 2-μm-thick tin–copper finish after nanoindentation tests after 3000 s of dwell time. Whiskers were found with the application of more than or equal to 4 mN [20].

are 1, 2, 4, 8, and 80 mN. During the holding process, the increasing indentation displacements due to creep is recorded continuously. Dwell time is 300 or 3000 s. For each test condition, tests were carried out three more times to check the repeatability.

Figure 9.8 shows indented surfaces at 1, 2, 4, and 8 mN for 2-μm-thick tin–copper finish. Contact area increases with the application of the indenter's load. No whisker was observed when the applied load is less than 2 mN. Whiskers were found at more than or equal to 4 mN [20]. It implies that there is a threshold load to produce tin whiskers by mating pressure. Whiskers formed at the middle of the triangular indented contact edge because of the indenter's shape. The diameter of whiskers is from 200 to 300 nm. When the applied load is 8 mN, the number of whiskers increases. Some whiskers are compressed by the indenter's face.

Figure 9.9 shows pressure-induced tin whiskers on the matte tin finish. The thickness of the finish is 5 μm with an applied load of 4 mN. Whiskers are formed near the contact area, but away from the edge of the contact. The diameter of the whiskers is almost same as the grain size. Whisker size depends on the grain size of the finish.

Table 9.2 shows the summary of nanoindentation tin whisker tests. The number of defects (nodules, hillocks, or whiskers) was also counted along the indented area. As the grain size increases, the number of defects is prone to increase. The minimum load to produce tin whiskers is defined when defects are found along the indented area with the application of the lowest load. The minimum load of tin–copper is higher than those of the other two finishes. This means that the grain size affects the minimum

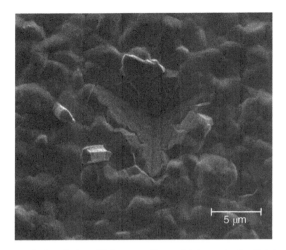

Figure 9.9 Pressure-induced tin whiskers on 5-μm-thick matte tin finish. Applied load is 4 mN [20].

TABLE 9.2   Summary of Nanoindentation Tin Whisker Tests for Different Finishes

Finish Type	SnCu	Bright Tin	Matte Tin
Grain size	Fine	Small	Big
Observed defects	Filament/Nodule	Nodule	Columnar
Number of defects	Few (2 μm)	Many (2 μm)	Many (2 μm)
	Few (5 μm)	Few (5 μm)	Many (5 μm)
Minimum load (mN)	4 (2 μm)	1 (2 μm)	1 (2 μm)
	4 (5 μm)	4 (5 μm)	2 (5 μm)

load to produce defects such as tin whiskers. As the thickness of the finish increases, the minimum load tends to be higher. In this experiment, since indented depth often reached over 1 μm, the substrate can affect the behavior of stress evolution when the finish thickness is 2 μm.

9.3.3   Growth Model

Fisher et al. [3] concluded that the rate of growth is proportional to the applied pressure. Atom transport is thought to be required to form whiskers [3, 18, 20–22]. Theoretically, a chemical potential on a grain boundary can be expressed as

$$\mu = -\sigma \Omega \tag{9.6}$$

Here, σ is a normal stress on the grain boundary and Ω is an atomic volume. Atomic diffusion is driven by the difference in the chemical potential and atomic flux. For

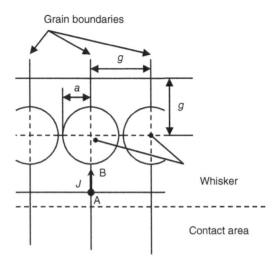

Figure 9.10   Pressure-induced tin whisker growth model.

spontaneous whisker growth, Tu has proposed a whisker growth model based on a constant biaxial compressive stress and grain boundary diffusion of tin [22].

Here, pressure-induced tin whisker growth is modeled as shown in Figure 9.10. A plated finish is assumed to consist of square columnar grains with a size of g. A pressure by a contact between components is applied on the surface of the plating, and σ_0 is defined as the stress at the point A (triple junction). Whiskers with a diameter of $2a$ are arrayed along the boundary of the pressured area periodically. Since the stress on the whisker bottom is negligible, compared with the applied stress, the difference in the chemical potential between points A and B can be approximated to be $-\sigma_0\Omega$. Atoms move along the grain boundary AB due to the difference in the chemical potential and the atomic flux along point A (J). Thus, J can be expressed as follows:

$$J = \frac{D_b \delta_b \Omega}{kT} \frac{\sigma_0}{g - a} \tag{9.7}$$

Here, D_b and δ_b are a grain boundary diffusion coefficient and a diffusion thickness, respectively, k is the Boltzmann constant, and T is the absolute temperature. Atoms are transported through point A to form the whisker. The growth rate of whisker length, v, is proportional to the stress at the point A, and thus,

$$v = \frac{J d_0}{\pi a^2} = \frac{D_b \delta_b \Omega \sigma_0 d_0}{\pi k T a^2 (g - a)} \tag{9.8}$$

where d_0 is the thickness of the plating.

9.4 CASE STUDIES

For electronics components, several factors affect the stress field of the finish: (i) thickness of the plating, (ii) structure of the ambient component material, and (iii) material microstructures. When the plating is thin, the volume "pushed out" by creep is small and the stress level is low. When the plating is subjected to pressure while constrained by the substrate, the stress level is higher.

The shape (structure) of the connector characterizes the stress distribution in the plating. A sharp-edged connector causes a higher stress concentration below a pressured surface. On the other hand, a larger radius of the connector mitigates the constraint of deformation in the plating near the pressured area and lowers residual stresses.

The local stress distribution is also affected by the microstructure of the plating. Since IMC-induced stresses are affected by the microstructure, spontaneous whiskers strongly depend on the nature of the microstructure. As a result, spontaneous tin whisker formation often exhibits random behavior. On the other hand, pressure-induced whiskers are usually formed near the pressured area. This shows that microstructure-induced stress is not the primary driving force.

Whisker formation is a stress relief phenomenon, and, therefore, best experimental practice requires that stresses be measured prior to whisker initiation. Since several factors affect the behavior of stress evolution in electronics components, the stress evolution should also be measured when assessing pressure-induced whisker formation. One experimental approach is XRD [23] to assess stress in a small area of the plating. However, for spontaneous whiskers, it is difficult to measure stress evolution by XRD because they grow randomly. For pressure-induced whiskers, such a measurement is complicated by the large stress gradient.

An alternative approach to measurement is finite element analysis (FEA) [4, 24, 25]. From each set of material properties and the configuration of each component, the stress field at any point can be calculated. In particular, using the creep property of plating, the stress evolution in the plating can be assessed. From a macroscopic view, contact analysis similar to the indentation to a creep body has been verified based on experimental data. This section presents several case studies with the FEA approach to assess the stress evolution in the plating and then to relate this to pressure-induced whisker formation.

9.4.1 Mating and Ball Indentation Tests

To assess pressure-induced tin whiskers in electronics components, several test methods have been proposed [4–6]. The mating test shown in Figure 9.11a is the simplest. A paired connector and cable are mated and then exposed to environmental conditions. Since the components' shapes affect the test results, the test results are not easily generalized. Another test method is the ball indentation test shown in Figure 9.11b. Here, a rigid ball is pressed onto a flat area on the connector side.

JEITA has performed the aforementioned two tests for various finishes. Their reports show good correspondence between the two tests. However, in some cases,

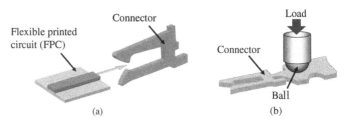

(a)   (b)

Figure 9.11 Test methods for pressure-induced tin whisker. (a) Mating test. (b) Ball indentation test.

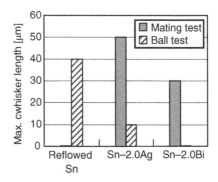

Figure 9.12 Pressure-induced tin whisker tests for reflowed tin, Sn–2Ag, and Sn–2Bi showing maximum tin whisker lengths in the mating test and the ball indentation test reported. Source: JEITA [5, 6].

there are inconsistencies between them. Figure 9.12 shows the experimental results reported by JEITA [6]. Three finishes – reflowed tin, Sn–2Ag, and Sn–2Bi – were assessed in both tests. The mating test was carried out at room temperature. Observations were conducted after 1000 h. A gold-finished flexible printed circuit (FPC) was used to avoid whiskers from the side of the FPC. The thickness of the FPC was about 260 µm, and a contact force was designed to be 1 to 2 N. The ball indentation test was also carried out at room temperature. A ZrO_2 ball ($\phi = 1$ mm) was used as an indenter. Tests were carried out with the application of 200 g. After 240 h of dwell time, the maximum length of whiskers was observed. The mating test results showed that reflowed tin grew no pressure-induced tin whiskers. However, during the ball indentation test, the longest whisker was found on the reflowed tin. Cross-sectional analysis showed the difference in thickness of each finish [6]. Table 9.3 shows the thickness of the plating and of the nickel underlayer, as reported by JEITA.

Figure 9.13 shows the creep properties of reflowed tin, Sn–2Ag, and Sn–2Bi. Three curves were based on Equation 9.1, and creep parameters were obtained from the nanoindentation test. In this graph, the slope of the curve is equivalent to the creep exponent in Equation 9.1. Although the creep exponents for the three materials range from 8 to 10, there is a large difference in the creep strain rate because of the creep

TABLE 9.3   Thickness of Plating and Nickel Underlayer for Reflowed Tin, Sn–2Ag, and Sn–2Bi

Sample	Thickness	
	Plating (μm)	Ni (μm)
Mating test		
Reflowed tin	0.5	0.5
Sn–2Ag	1.7	0.5
Sn–2Bi	1.5	0.5
Ball indentation test		
Reflowed tin	1.2	0.3
Sn–2Ag	0.3	0.7
Sn–2Bi	0.7	0.3

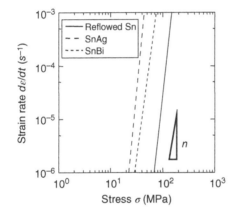

Figure 9.13   Creep properties of reflowed tin, Sn–2Ag, and Sn–2Bi finishes.

constant. The creep properties suggest that Sn–2Ag and Sn–2Bi should be more resistant to pressure-induced whisker formation than is reflowed tin. However, this is not borne out by the aforementioned experimental results. This inconsistency suggests that the stress evolution must also be considered when predicting pressure-induced tin whisker behavior.

FEA models were constructed for the ball indentation test and the mating test. Figure 9.14a shows the FEA model and boundary conditions for ball indentation. An axisymmetric model was used to model the plating, underlayer, and substrate. The thickness of the plating and underlayer was taken from Table 9.3. The applied load and dwell times were 200 g and 240 h, based on the experimental conditions [6].

To model the mating test, the connector was simplified to a ball as shown in Figure 9.14b. The plating and underlayer were modeled on the ball. On the opposite side (FPC side), the model consists of a gold plating (1 μm), nickel underlayer

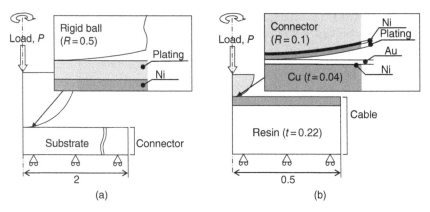

Figure 9.14 FEA models for (a) ball indentation and (b) mating tests (unit: mm).

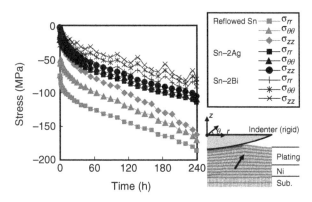

Figure 9.15 FEA qualification of stress evolution in the plating for the ball indentation test.

(0.3 μm), copper layer (40 μm), and resin (220 μm). The thickness of each layer was taken from the JEITA report [6]. An applied load was set to 200 g as an approximated value. The dwell time was 1000 h.

In both FEA tests, the commercial solver MSC.MARC was used and a four-node two-dimensional element was used for mesh division. Fine mesh was constructed in the plating layer and minimum mesh size was 0.1 μm. Radial displacements along the symmetric axis and vertical displacements on the bottom of the substrate were fixed. Friction on the contact surface was not considered. The plating was assumed to be an elastic–plastic creep body. Young's modulus of the plating was obtained from the unloading segment of the indentation test with Poisson's ratio of 0.3. Yield stress was 1/3 of the hardness from the nanoindentation test. The hardness was taken at the beginning of holding time. Other materials are assumed to be isotropic elastic bodies.

Figure 9.15 shows the FEA results for the ball indentation test. In the graph, σ_{rr}, $\sigma_{\theta\theta}$, and σ_{zz} of each plating material are plotted. These values are extracted at the

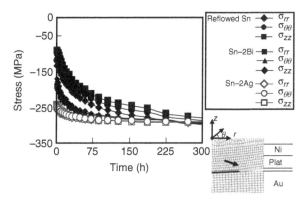

Figure 9.16 FEA qualification of stress evolution in the plating for the mating test.

middle point of each finish shown in the figure. The components of stress increase as a function of time even though stress relaxation takes place. At 240 h, reflowed tin has the highest compressive stresses and is expected to have longer whiskers. According to the experimental results, the length of whiskers in reflowed tin is 40 µm and longer than others. On the other hand, no whiskers were found on the Sn–2Bi having the lowest stress level.

Figure 9.16 shows the FEA results for the mating test. In contrast to Figure 9.15, Sn–2Ag has the highest stress level in the early stage. According to the JEITA report, whiskers were found on Sn–2Ag just after 3 h and on Sn–2Bi after 120 h [5]. The longest whisker was found on Sn–2Ag plating from Figure 9.12. The FEA results agree with the experimental data. These two analyses show that higher compressive stresses cause longer and more rapid whisker formation.

To evaluate the growth rate in the ball indentation test shown by Equation 9.8, we take $D_b\delta_b = 6 \times 10^{-22}$ m^3/s, $\Omega = 2.7 \times 10^{-29}$ m^3, and $T = 298$ K [22]. Based on experimental observations, whisker size ($2a$) and grain size (g) are approximated to be $2a = g = 3$ µm. The σ_0 is taken from FEA and d_0 is taken from Table 9.3. The growth rate of the pressure-induced whisker is shown in Figure 9.17. In this prediction, no incubation time is considered. After 240 h in the ball test, the estimated maximum whisker length of reflowed tin and Sn–2Ag are 55 and 8 µm, respectively. These values agree with the experimental results. Since several other factors (incubation time, effect of grain boundary and/or IMC networks, etc.) are not considered, this growth model can be improved. However, this model enables prediction of the whisker growth in the electronics components.

9.4.2 Effect of Connector's Shape on Loading Test

A tip of the connector is usually designed based on contact resistance. Therefore, the shape of the tip changes for the purpose of different products. Since the shape of tip affects the local stress distribution of contact area, whiskers also depend on the connector's shape. Figure 9.18 shows a schematic illustration of loading test with

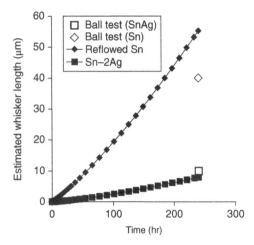

Figure 9.17   Prediction of maximum whisker length in the ball indentation test.

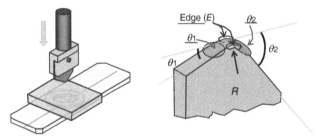

Figure 9.18   Effect of connector's shape on loading test. (a) Loading test with connector's tip. (b) Magnification of connector tip. Please visit www.wiley.com/go/Kato/TinWhiskerRisks to access the color version of this figure.

several connector tips. The shape of the connector is determined by curvature (R) and angles of two edges (θ_1 and θ_2).

Six shapes were assessed as shown in Figure 9.19. The angle of one edge surface (θ_1) is 15° or 30°, and the angle of another edge surface (θ_2) is kept at 60°. The radius of curvature (R) is 0.1, 0.3, or 0.5 mm. The value of load is 200 g, and test time is 120 h. Observations were carried out at three areas; two edge surfaces (θ_1 and θ_2) and the edge line on the side of connector (E). Whisker length depended on the curvature of the tip, R. The stresses near the contact area are sensitive to the curvature of the connector's tip. Longer whiskers were found at the area of the side edge (E). Since the side edge is sharp, severe stresses are generated along the side edge.

9.4.3   Flexible Printed Circuit (FPC)

FPC consists of surface finish, copper film, substrate, and support. Copper/substrate and substrate/support are bonded with adhesives. Since adhesives are softer than other

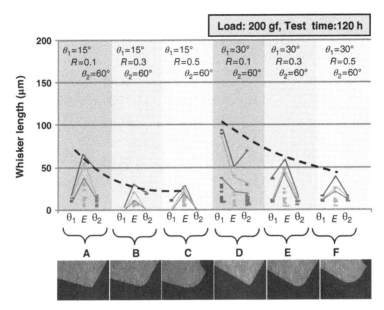

Figure 9.19 Whisker growth on loading test with six connector tips. Please visit www.wiley.com/go/Kato/TinWhiskerRisks to access the color version of this figure.

materials, deformation behavior of separable contacts on FPCs is affected by the mechanical properties of adhesives. Figure 9.20 shows an example of a Vickers indentation test. The FEA result is also shown. The shape of the contact area is different from that of the Vickers indenter's tip due to adhesives. Many nodules are found at the highly stressed area.

9.4.4 Double-Layered Tin–Copper Plating

Double-layered tin–copper plating is semibright tin–copper over matte tin–copper. It was reported that a connector with double-layered tin–copper plating shows good results in mating tests, but the mitigating mechanism was not clear [26]. Table 9.4 shows the material properties of semibright (layer I) and matte (layer II) tin–copper finishes obtained from the nanoindentation creep test for a connector with double-layered tin–copper plating. In Table 9.4, the creep property of the layer II was taken from an approximated value where the tip indented with 4 mN can penetrate the layer I. Total thickness of layers I and II is 5 μm. The creep rate of semibright tin–copper is five times faster than that of matte tin–copper.

Figure 9.21 shows the stress evolution on double-layered plating for the ball indentation test. The plating consists of layers I and II; the fraction of layer II is set to 80%. Curves show the axial components of stress at the middle of layer II on the contact edge. After 24 h, stresses have significantly decreased. The results for single layer II are also plotted. In the case of a single layer, an increase in stresses can be seen. This

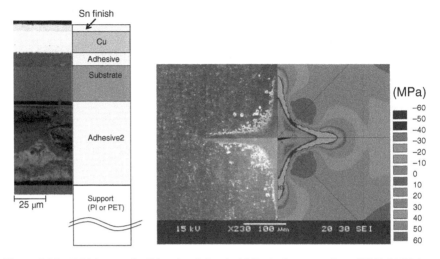

Figure 9.20 Whiskers on flexible printed circuit. (a) Typical cross section of FPC. (b) Vickers indentation test on FPC. Please visit www.wiley.com/go/Kato/TinWhiskerRisks to access the color version of this figure.

TABLE 9.4 Mechanical Properties of Semibright and Matte Tin–Copper

	Creep Exponent n	Creep Constant A (MPa^{-n})
Semibright SnCu	4.7	3.4×10^{-11}
Matte SnCu	4.5	1.5×10^{-10}

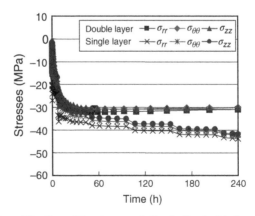

Figure 9.21 FEA qualification of the stress evolution in the double-layered tin–copper plating for the ball indentation test.

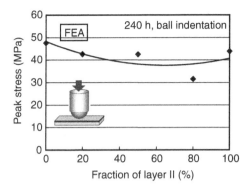

Figure 9.22 Effect of the fraction of matte tin–copper layer on the stress evolution.

shows that double-layered structure mitigates high multiaxial residual stresses and should thereby also mitigate the pressure-induced whisker formation.

In Figure 9.22, the vertical axis indicates the peak stress evolution from Figure 9.21. Interestingly, the peak stress at 80% of the fraction is the minimum, and the mating test showed that no whiskers were found when the fraction of layer II was 80%. Clearly, stress evolution again corresponds to pressure-induced tin whisker formation.

9.5 CONCLUSIONS

Pressure-induced tin whiskers were discussed in this chapter. Since pressure-induced tin whiskers grow rapidly and can bridge adjacent conductors, there is a serious risk to separable connectors. To evaluate the whisker length quantitatively, a pressure-induced tin whisker growth model was presented. Based on nucleation theory, a threshold stress is required for tin whisker growth. The critical stress is related to whisker diameter. The growth model based on grain boundary diffusion theory can predict the maximum whisker lengths.

As a methodology for the assessment of pressure-induced tin whisker formation, the creep-based tin whisker formation model shows good results. Several connectors with tin or tin-rich alloy finishes were assessed by means of the nanoindentation technique. Measured creep properties correspond to mating whisker tests for connectors and FPCs. In particular, the creep exponent characterizes the deformation mechanism of the plating and shows that the pressure-induced whisker formation depends on the creep mechanism. A higher creep exponent generates higher stresses because stress relaxation occurs more rapidly. In fact, the longest whiskers were found on Sn–1.5Cu plating with the highest creep exponent, and the shortest ones were found on the Sn–10Pb with the lowest creep exponent. However, since the creep property of the plating depends on the manufacturing process, the creep exponent should be measured for each assessment. The creep property then can be a tin whisker risk screen for plating finishes.

Since the assessment of electronics components depends on the components and the plating process, stress evolution using FEA was conducted. According to the JEITA report [6], the ball indentation tests for reflowed tin, Sn–2Ag, and Sn–2Bi contradicted the mating test for the same samples because of the difference in thickness of the plating at assessed points. However, stress evolution obtained from FEA showed a good correspondence to each experimental result. Higher stress evolution corresponds to longer whisker growth. Several case studies were shown in this chapter. Effect of the connector's tip on the loading test was also shown. The curvature of the connector's shape is a key factor for mechanically induced tin whisker formation. For the FPC, mechanical properties of adhesives are key factors influencing tin whisker formation. Each mechanical property can be obtained by the indentation technique, and stress distribution can be estimated by FEA.

REFERENCES

1. Ganesan S, Pecht M. *Lead-free Electronics*. Hoboken, NJ, USA: John Wiley & Sons, Inc.; 2006.

2. Leidecker H, Brusse J. Tin whiskers: A history of documented electrical system failures. Technical Presentation to Space Shuttle Program Office; NASA, April 2006.

3. Fisher RM, Darken LS, Carroll KG. Accelerated growth of tin whiskers. Acta Metall 1954;2:369–373.

4. Shibutani T, Yu Q, Shiratori M, Pecht MG. Pressure-induced tin whisker formation. Microelectron Reliab 2008;48:1033–1039.

5. JEITA, ed., Annual Report of Lead-Free Solder for Practical Use 2004; JEITA. pp 1–32, 2004.

6. JEITA, ed., Annual Report of Lead-Free Solder for Practical Use 2005; JEITA. pp 11–22, 2005.

7. Vo N, Kwoka M, Bush P. Tin whisker test standardization. IEEE Trans Electron Packag Technol 2005;28(1):3–9.

8. Sakamoto I. Whisker test methods of JEITA whisker growth mechanism for test methods. IEEE Trans Electron Packag Technol 2005;28(1):10–16.

9. Pitt C, Henning R. Pressure-induced growth of metal whiskers. J Appl Phys 1964;35:459–460.

10. Yang F, Li Y. Indentation-induced tin whiskers on electroplated tin coatings. J Appl Phys 2008;104:113512.

11. Lin CK, Lin T-H. Effects of continuously applied stress on tin whisker growth. Microelectron Reliab 2008;48:1737–1740.

12. Chen YJ, Chen C-M. Mitigative tin whisker growth under mechanically applied tensile stress. J Electron Mater 2008;38(3):415–418.

13. Southworth AR, Ho CE, Lee A, Subramanian KN. Effect of strain on whisker growth in matte tin. Solder Surf Mt Technol 2008;20(1):4–7.

14. Moriuchi H, Tadokoro Y, Sato M, Furusawa T, Suzuki N. Microstructure of external stress whiskers and mechanical indention test method. J Electron Mater 2007;36(2):220–225.

15. Mayo MJ, Nix WD. A micro-indentation study of superplasticity in Pb, Sn, and Sn–38 wt.% Pb. Acta Metall 1988;36(8):2183–2192.

16. Takagi H, Dao M, Fujiwara M, Otsuka M. Experimental and computational creep characterization of Al–Mg solid-solution alloy through instrumented indentation. Philos Mag 2003;83(35):3959–3976.

17. Shibutani T, Yu Q, Shiratori M. A study of deformation mechanism during nano-indentation creep in tin-based solder balls. J Electron Packag 2007;129:71–75.

18. Boettinger WJ, Johnson CE, Bendersky LA, Moon K-W, Williams ME, Stafford GR. Whisker and Hillock formation on Sn, Sn–Cu and Sn–Pb electrodeposits. Acta Mater 2005;53(19):5033–5050.

19. Lee B-Z, Lee D-N. Spontaneous growth mechanism of tin whiskers. Acta Metall 1998;46(10):3701–3714.

20. Shibutani T. Effect of grain size on pressure induced tin whisker formation. IEEE Trans Electron Packag Technol 2010;33(3):177–182.

21. Tu KN. Interdiffusion and reaction in bimetallic Cu–Sn thin films. Acta Metall 1973;21(4):347–354.

22. Tu KN. Irreversible processes of spontaneous whisker growth in bimetallic Cu–Sn thin film reactions. Phys Rev B 1994;49(3):2030–2034.

23. Choi WJ, Lee TY, Tu KN, Tamura N, Celestre RS, Mcdowell AA, Bong YY, Nguyen L, Sheng GTT. Structure and kinetics of Sn whisker growth on Pb-free solder finish. In: Proceedings 52nd Electronic Components and Technology Conference; 2002 May 28–31; IEEE. pp 628–633.

24. Zhao J-H, Su P, Ding M, Chopin S, Ho PS. Microstructure-based stress modeling of tin whisker growth. IEEE Trans Electron Packag Technol 2006;29(4):265–273.

25. Lau JH, Pan SH. 3D nonlinear stress analysis of tin whisker initiation on lead-free components. J Electron Packag 2003;125:621–624.

26. Shibutani T, Wu J, Yu Q, Pecht M. Key reliability concerns with lead-free connectors. Microelectron Reliab 2008;48(10):1613–1627.

INDEX

AATC *see* air-to-air thermal cycling test (AATC)
activator, 219, 222
activity, 219–222
adipic acid, 219
Ag_3Sn, 153
air, 36, 217, 219, 222
air-to-air thermal cycling test (AATC), 43, 44, 51
algorithm, 56, 175, 177–180, 182, 185
alloy 42, 72–74, 76, 78–81, 126, 127, 129, 130, 135, 142, 165, 219, 220
alloying, 69–71, 75–89, 118, 164
ambient, 76, 78, 82–84, 87, 109, 165, 188, 195, 197–200, 202–206, 208–213, 231, 237
anisotropy, 2, 13–17, 44–46, 49, 50, 60, 64, 66, 144
annealing, 4, 166, 197
ANSI, 185
aspect ratio, 76, 189
assembly, 70, 72, 75, 161–162, 164, 174, 178, 179, 183, 184, 217–218, 220–221
ATC, 215, 216, 220
atomic diffusion, 110, 235
atomic flux, 5, 40, 107–109, 235, 236
atom migration, 229
atom transport, 226, 235

axis, 3, 46, 48, 62, 63, 65, 139, 140, 150–152, 198, 240

bake, 166
ball indentation, 237–246
barrier, 83, 109, 162, 165, 176, 177
β-Sn, 15, 17, 44–46, 50, 58, 102, 128, 146, 148, 149, 153, 154
bi-axial modulus, 204
bismuth (Bi), 70, 71, 73, 75, 80, 81, 84, 89, 165, 231
Boltzmann constant, 236
brass, 165, 166, 188, 194–202, 205, 208–211, 219
bridge, 162, 169, 171, 173, 178, 179, 226, 245
bright, 25, 26, 82, 87, 89, 95, 99, 126, 128, 129, 188, 194, 195, 197, 199, 202–205, 209–212, 232, 233, 235
bright field, 95, 99, 126, 128, 129

C151, 220
C194, 72, 76, 78–80, 219
C7095, 219
C18045, 90, 117, 118

Mitigating Tin Whisker Risks: Theory and Practice, First Edition.
Edited by Takahiko Kato, Carol A. Handwerker, and Jasbir Bath.
© 2016 John Wiley & Sons, Inc. Published 2016 by John Wiley & Sons, Inc.
Companion website: www.wiley.com/go/Kato/TinWhiskerRisks

C19400, 90, 117, 118, 130
CALCE *see* center of advanced life cycle
 engineering (CALCE)
carbon, 126, 130
CDA number, 90, 117, 118, 130
center of advanced life cycle engineering
 (CALCE), 168
chemical potential, 107, 235, 236
chloride (Cl⁻), 219, 222
chlorine, 131
chromium (Cr), 90, 95
coating, 58, 62, 63, 89–91, 93–109, 111–118,
 126–128, 130, 162–164, 167–173,
 176–179, 181, 216, 217, 225
coble creep, 4, 7, 17, 30, 196
coefficient of thermal expansion (CTE), 12, 13, 15,
 44, 45, 50, 59, 106, 165, 178, 179, 198, 203
columnar, 24, 30, 31, 33, 55, 56, 58, 59, 82, 93,
 105, 149, 151, 152, 154, 226, 231, 236
commercial off-the-shelf (COTS), 173–175, 184
compressive strain, 13, 203, 208, 228
compressive stress, 6, 12, 17, 23, 29–31, 38, 43,
 46, 59, 60, 65, 66, 80–82, 86–89, 102–105,
 107, 109, 115, 116, 127, 142, 208, 227, 228,
 232, 233, 236, 241
conformal coat, 162–172, 176–179, 181
connector, 135, 141, 152, 225–227, 230, 231,
 237–243
 shape, 241–242, 246
 tip, 242, 246
contact
 area, 226, 234, 241–243
 edge, 234, 242
 finish, 225, 226
 forces, 226, 238
contamination, 133, 165, 179, 219, 220, 222
copper (Cu), 46, 57, 70, 72, 76, 78–80, 82–86, 89,
 90, 95, 96, 98, 118, 127, 128, 130, 131, 133,
 165, 166, 198, 219, 226, 227, 229, 240, 242
copper–chromium (Cu–Cr), 95
copper–tin (Cu–Sn), 25, 81, 82, 86, 102, 165, 196
corrosion, 23, 78, 127, 128, 131, 163, 198, 225,
 226
COTS *see* commercial off-the-shelf (COTS)
crack, 6, 7, 12, 35
cracked grain, 35, 36, 203
creep
 constant, 229, 244
 exponent, 230, 231, 238, 244, 245
cross-section, 3–8, 24, 36, 80, 82–84, 86, 87, 90,
 93–101, 104, 105, 112, 116–118, 127–132,
 149–154, 189, 192, 195, 203, 238, 244
crystalline orientation, 93–95, 117, 126
crystallography, 2

Cs-corrector, 126
CTE *see* coefficient of thermal expansion (CTE)
cubic, 46, 141
curved, 4, 6, 135
Cu₃Sn, 126, 128
Cu₆Sn₅, 2, 13, 25, 82, 87, 95, 96, 98, 102–104,
 109, 126, 128, 129, 142
Cu₇Sn₅, 96, 104, 128
(Cu,Ni)₆Sn₅, 153, 154

dark field, 128–132
dark field-STEM (DF-STEM), 131, 132
defect density, 87, 192–194, 199, 202, 205
defect length, 189
defect volume, 188, 194, 198, 199, 203, 205, 208,
 209
deflection, 232
deformation, 2, 23, 24, 38, 39, 50, 80, 105, 139,
 151, 226, 229, 230, 237, 243, 245
ΔT, 198
deposition, 4, 24, 25, 27, 35, 36, 39, 89, 126, 163,
 179, 195
designer, 161–163, 173, 174
DF-STEM *see* dark field-STEM (DF-STEM)
diffraction, 31, 58, 126, 127, 134, 135, 149, 155
diffraction angle, 58
diffusion, 2, 4, 12, 23, 27, 28, 30–34, 36, 37, 39,
 48, 80, 82, 83, 90, 96, 98, 100, 101, 104, 107,
 109, 110, 118, 226, 228, 235, 236, 245
diffusion coefficient, 110, 111, 236
diffusion-induced grain boundary migration, 9
dislocation, 4, 82, 141
displacement, 110, 229, 234, 240
driving force, 12, 13, 21–40, 43, 46, 65, 66, 195,
 196, 237
dwell time, 234, 238–240

EBSD *see* electron backscatter diffraction (EBSD)
edge line, 242
edge surface, 242
EDPs *see* electron diffraction patterns (EDPs)
EDS *see* energy dispersive X-ray spectroscopy
 (EDS)
EDX *see* energy dispersive X-ray spectroscopy
 (EDX)
EELS *see* electron energy-loss spectroscopy
 (EELS)
elasticity anisotropy, 44–46
elastic lattice strain field, 109
elastic–plastic deformation, 105
elastic strain, 2, 15, 17, 18, 196, 198, 208
elastic strain energy density (ESED), 3, 14–16,
 196, 198

electrodeposition, 24, 86, 90, 91, 98, 99, 102, 118, 126, 127, 129–131
electromigration, 22, 23
electron backscatter diffraction (EBSD), 2, 90, 125, 144–147, 155, 195
electron backscatter diffraction pattern (EBSP), 90, 93, 94, 112, 114–116, 147–149, 155
electron beam, 24, 95, 126–128, 135, 148
electron diffraction patterns (EDPs), 126, 134–135
electron energy-loss spectroscopy (EELS), 126
electroplated, 10, 28, 87, 89, 192, 194–195, 198, 203–209
embrittlement, 80
end user, 164, 175
energy dispersive X-ray spectroscopy (EDS), 183
energy dispersive X-ray spectroscopy (EDX), 94–101, 126, 130, 131, 133
equiaxed, 24, 30, 82, 86
ESED *see* elastic strain energy density (ESED)

fatigue, 10–13
FEA *see* finite-element analysis (FEA)
FEM *see* finite element method (FEM)
FeSn$_2$, 126, 135
FGIMC layer *see* fine-grained intermetallic compound (FGIMC) layer
FIB *see* focused ion beam (FIB)
field-emission scanning transmission electron microscopy (FE-STEM), 90, 93–95, 99, 105, 118, 128–131
field-emission transmission electron microscopy (FE-TEM), 90, 93–95, 126–129
filamentary, 36, 82, 87, 91, 135, 141, 226, 231, 232
film
    formation, 226
    microstructure, 3, 188, 209
    thickness, 4, 26, 27, 59, 82, 192, 195, 198, 203, 204, 209
fine-grained intermetallic compound (FGIMC) layer, 96–101, 104, 105
finite-element analysis (FEA), 23, 31–34, 36, 38–40, 90, 105–109, 115–118, 237, 239–241, 243–246
finite element method (FEM), 44–46, 50, 55, 56, 58–60
flux, 5, 6, 12, 17, 34, 40, 72, 107–109, 217, 219–220, 235, 236
focused ion beam (FIB), 2–6, 23, 24, 34, 40, 50, 51, 93, 112, 126, 129–133, 191, 203
foreign object debris (FOD), 176, 177
free surface, 82, 189, 208
friction, 6, 7, 13, 232, 233, 240

GEIA, 174, 185
geometry, 2–4, 6–8, 13, 17, 50, 55–58, 161–163, 167, 172, 173, 179, 196
gold (Au), 131, 132
grain boundary, 1–18, 31–34, 36, 37, 39, 49, 82–84, 87–89, 97, 106, 107, 109, 110, 143, 189, 196, 231, 233, 235, 236, 241, 245
grain boundary diffusion, 30–34, 36, 39, 82, 83, 236, 245
grain diameter, 93, 197
grain growth, 4, 23, 36, 37, 39, 40, 189
grain orientation, 14, 15, 44, 46–50, 55, 58, 61, 93, 117, 215–216
grain rotation, 44, 52–55
grain size, 24–29, 44, 56, 59, 93–95, 148, 149, 195, 203, 215, 216, 230, 232, 234, 235, 241

hardness, 229, 240
HBr, 219
hexagonal, 31, 96, 140
hillock, 2–17, 187–213, 232, 234
histogram, 193, 198, 200, 204, 206, 208
hkl, 13, 48, 60, 61, 198
Hooke's law, 44
H$_2$S, 226
humidity, 76, 78, 80, 83, 84, 87, 127, 131, 188, 195–198, 203, 216, 219, 220, 222, 226

IMC volume, 24–29
incubation time, 27, 35, 241
indentation, 229, 230, 233, 237–246
indented area, 234
indented depth, 235
in-plane stress, 25, 26, 228
intermetallic compounds (IMC), 3, 22, 43, 81, 95, 96, 126, 128, 135, 142, 165, 196, 218
International Electronics Manufacturing Initiative (iNEMI), 78, 188, 194, 209
International Electrotechnical Commission (IEC), 76, 78
inverse pole figure (IPF), 14–16, 93, 149
iron (Fe), 95
irregular growth, 190, 202, 209
isothermal, 3, 4, 195

Japan Electronics and Information Technology Industries Association (JEITA), 72, 74, 76, 82–84, 118, 166, 231, 237, 238, 240, 241, 246
Joint Electron Device Engineering Council (JEDEC) standards, 44, 76, 78, 160, 183–185, 188, 189, 191, 195, 216, 217

Kikuchi line, 148, 149
kink angle, 135, 141, 143–145
kinked, 140, 145, 151, 189
kinks, 36, 141, 143–145, 150
Kirkendall, 98, 99, 101, 128

large-grained intermetallic compound (LGIMC),
    95–102, 104–107, 114–118
lattice, 83, 109–111, 131, 135–144, 148, 155
lauryl amine, 219
lead (Pb), 69, 81, 164
lead-frame, 44, 59, 89–118, 165, 218–221
lead-free (Pb-free), 69–73, 75, 76, 78, 81, 83, 135,
    164, 216–220, 222, 225–227
lead–tin (Pb–Sn), 24, 29, 30
LGIMC see large-grained intermetallic compound
    (LGIMC)
load, 51, 60, 63, 106, 227–230, 233–235, 239,
    240
low-profile quad-flat-package (LQFP), 72

mating test, 226, 232, 234, 237–241, 243, 245,
    246
matte, 70, 76, 82, 89–118, 175, 177, 182, 188,
    194, 195, 197, 199, 202–205, 207, 209–213,
    220, 226, 230–235, 243–245
mean time between failures (MTBF), 175
mechanical contact stress, 83, 86, 87
mechanical stress, 21, 22, 135, 141, 149, 152–154
melting point, 46, 71–73, 216–218, 229
metal finisher, 159–161
metastable, 136
micropillar, 133–135
microstructure, 1–18, 23, 24, 27, 29, 30, 33, 40,
    55, 90, 93, 95, 105, 109, 111, 112, 114,
    116–118, 147, 188, 190, 196, 197, 203, 209,
    216, 230, 237
migration, 2–5, 8–10, 163, 189, 229
Mil-STD, 183
minimum load, 234, 235
misorientation, 13–18, 150, 151
mitigation, 21, 75–89, 118, 126, 135, 155,
    159–185
mitigator, 162
modeling, 1–18, 21, 31–34, 38, 40, 43–66, 88,
    102–106, 110, 111, 168–173, 175, 185, 195,
    196, 228–233, 235–236, 239–241, 245
moisture sensitivity, 216
molecular dynamics, 9, 90, 109–111, 115, 118
monoclinic, 142
Monte Carlo, 170, 177
morphology, 4, 24, 34, 36, 39, 87, 96, 98–104,
    194, 200, 201, 206, 207, 209

MUD, 204
multibeam optical stress sensor (MOSS), 25
multiple grain, 55–66

nano-diffraction, 127–129
nanoindentation, 229, 230, 233–235, 238, 240,
    243, 245
National Institute of Standards and Technology
    (NIST), 165, 183, 187, 195
NiAu, 161
nickel (Ni), 72, 180, 226, 238–241
nickel–iron (Ni–Fe), 165
nickel–tin (Ni–Sn), 165
nickel underlayer, 165, 228, 229, 238, 239
NiPdAu, 161
Ni3Sn4, 126, 135
nitrogen, 219, 220
nodule, 153, 154, 234, 235, 243
noncorroded, 127, 128, 131
non-single crystal whisker, 111, 116
normal direction inverse pole figures (ND-IPFs),
    14–16
nucleation, 2, 11, 36, 38, 50, 51, 54, 245

object-oriented finite-element analysis, 13
odd shaped eruptions, 189, 191
optical microscope, 25, 131, 132, 231
orientation, 2, 4, 13–17, 38, 39, 44, 46–50, 55,
    58, 60, 61, 93, 106, 117, 126, 139, 147–155,
    198, 203, 204, 215–216
orthorhombic, 136, 138
oxidation, 22, 43, 81, 131, 143, 219, 222
oxide, 2, 4, 6, 7, 13, 17, 23, 24, 30, 31, 34–36,
    39, 40, 82, 91, 127, 130, 131, 196, 198, 208,
    226
oxygen, 43, 130

parylene, 163, 176, 177, 180
paste, 215, 217–222
paste volume, 215, 220–222
Pb-free, 176
peak temperature, 215, 217–222
phase field crystal (PFC) simulation, 9, 10
phosphor bronze, 87, 228, 232
phosphorous, 90
pinch-off, 10–13, 17
planar slicing, 90, 91, 109, 111–116, 118
plasticity, 23, 27, 29, 31, 32, 36, 39, 45, 50, 55, 60,
    64–66
plating
    chemistry, 160
    thickness, 164, 166, 227, 230
PLCC, 219, 220

Poisson's ratio, 59, 60, 102, 106, 229, 240
pole figure, 16, 114, 149–153
popped grain, 188–190, 200, 201, 205–209
potting, 162, 176, 177
power cycling, 43
PQFP, 216
preconditioning, 73, 76–81, 83, 84, 216, 217
preheat, 216, 218
pressure, 22, 23, 59, 226–229, 232–239, 241, 245
pressure-induced whisker, 226, 227, 229, 237, 239, 241, 245

radius of curvature, 242
recrystallization, 4, 5, 38
reflow
    atmosphere, 217, 219–222
    profile, 216–218, 220, 221
reflowed tin, 231, 238, 239, 241, 246
relaxation, 1, 17, 18, 23, 24, 27–35, 38, 39, 81, 82, 187–209, 229, 230, 241, 245
reliability, 2, 21, 43, 69, 71–73, 159–185, 218, 225–227
residual stress, 22, 23, 102, 103, 166, 229, 230, 232, 237, 245
residue, 217
ridges, 39, 87, 88, 169, 178, 179
risk management, 159–164, 173, 176–184
RoHS, 161, 182–184
ROL0, 220
rotational, 189

SAE, 166, 173, 175–177, 184, 185
satin, 188, 194, 195, 197, 199, 201–205, 207, 209–212
scanning electron microscopy (SEM), 3, 6, 9–11, 15, 16, 23, 25–28, 31, 32, 34–36, 39, 40, 51, 56, 76, 90–93, 111–116, 125, 126, 135–144, 148, 153, 183, 189, 190, 231
scanning ion microscope (SIM), 131, 132
scanning transmission electron microscopy (STEM), 125
SED see strain energy density (SED)
segmented length, 191
SEM see scanning electron microscopy (SEM)
semi-bright, 230, 231, 243, 244
shear yield, 233
shorting, 161, 169, 172, 188
silver (Ag), 12, 24, 30, 32, 69, 70, 73, 81, 86–87, 89, 134, 135, 165, 226, 231
silver–tin (Ag–Sn), 86
$\sin^2 \psi$ method, 102
$\sin^2 \psi - 2\theta$ diagram, 103
sliding, 3–10, 12, 13, 17, 18, 139, 196, 232

slip, 45, 46, 50, 54, 66
SMT, 162, 167
Sn^{2+}, 70, 71
Sn^{4+}, 71
Sn–Ag, 69–71, 73, 75, 76, 78, 80, 81, 84–87
Sn–2Ag, 74, 230, 238–242, 246
Sn–3.5Ag, 71, 77–80, 84
SnAgCu, 215, 216, 218, 220, 221
Sn–Ag–Cu, 72, 153, 154, 226
Sn3Ag0.5Cu, 218–220
Sn–2Bi, 77–80, 84, 230, 238–241, 246
Sn–58Bi, 71
Sn–Cu, 23, 24, 27, 31, 38, 39, 69–73, 75, 76, 78, 80, 81, 84, 86–109, 111–119, 128, 129, 141, 142, 149–154, 226, 235, 244
Sn–0.7Cu, 71
Sn–1.5Cu, 77–80, 84, 230, 245
Sn–3%Cu, 232, 233
Sn–2 mass%Cu, 130
SnO, 126, 127, 135, 136
SnO_2, 126, 127, 129, 135, 139
SnPb, 30, 43, 176, 177, 218, 221
Sn–Pb, 69–73, 75, 76, 80–87
Sn–2%Pb, 232
Sn–10Pb, 74, 77–80, 83, 230, 245
Sn37Pb, 220
Sn–Pb–Bi, 73
Sn+0.7wt%Cu, 16
SO_2, 226
SO_4^{2-}, 219
soak, 76, 218
SOIC, 172
solderability, 43, 70–73, 75, 160
solder dipping, 164, 166
solder joint, 70, 72–74, 163, 216–218
SOT, 219
standard deviation, 192–194
stencil aperture, 217, 220
stencil thickness, 217, 220
stereogram, 114, 116, 117, 149
strain, 2–4, 13, 15–18, 23, 29–31, 36, 38, 40, 44–46, 55, 59, 61, 63, 64, 105, 106, 109, 188, 189, 192, 196, 198, 199, 202–205, 208, 228, 229, 237–239
strain energy density (SED), 2, 3, 17, 38, 44, 46, 61, 63, 64, 196
strength, 3, 6, 70, 72–75, 196, 198, 203
stress
    concentration, 49, 50, 229, 237
    distribution, 33, 66, 90, 105–109, 118, 216, 237, 241, 246
    evolution, 23–34, 38, 39, 87, 235, 237, 239–241, 243–246

stress (*Continued*)
    gradient, 5, 6, 12, 13, 17, 31, 34, 38, 40, 106,
        107, 109, 116, 117, 237
    level, 38, 49, 165, 230, 237, 241
    relaxation, 1, 23, 24, 27, 29, 30, 32–35, 39, 81,
        82, 187–213, 229, 230, 241, 245
    relieved, 27, 29, 32, 43, 65, 237
    substrate, 4, 12, 13, 24, 25, 30, 109, 126–131, 135,
        142, 165–167, 178–181, 188, 194–213, 226,
        229, 230, 232, 235, 237, 239, 240, 242, 244
sulfate, 219, 222
sunken grain, 188–191, 195, 197, 203, 206–209
surface defect, 1, 3, 5, 7, 87, 187–213
surface free energy per unit area, 232
synchrotron, 15, 17, 18, 148
system, 2, 10, 23, 29, 38, 45, 46, 48, 50, 52, 55,
    60, 66, 93–95, 102, 109, 119, 126, 131, 133,
    135, 159–185, 196, 209, 225

TEM *see* transmission electron microscopy (TEM)
temperature, 3, 4, 10, 13–15, 17, 18, 30, 43, 44,
    46, 49, 55, 60, 63–66, 71, 74–81, 83, 84, 86,
    87, 90, 91, 106, 107, 109, 110, 118, 133, 141,
    165, 179, 188, 195, 197–199, 203, 205, 208,
    209, 214–222, 226, 229, 233, 236, 238
temperature and relative humidity, 195
temperature cycling, 43, 44, 55, 60, 63, 64, 66, 72,
    74, 76–80, 83, 86, 87
temperature/humidity, 76–80, 83, 84, 86, 87, 197,
    205
temperature/%relative humidity (T/RH), 197, 199,
    202–206, 210–212
tensile stress, 12, 27, 86, 107, 109, 165, 166, 208,
    228
terraces, 87, 88
test coupon, 192, 216, 217
tetragonal, 44, 50, 94, 135–137, 140, 146, 149
texture, 2, 3, 13–15, 17, 18, 24, 29, 44, 46, 48, 58,
    60, 61, 93, 195, 196, 198, 203, 204
thermal cycling, 3–5, 10–15, 17, 18, 22, 23, 43,
    55, 141–143, 188, 195, 197, 199, 200, 202,
    203, 205, 206, 208–210, 212, 220
thermoelastic, 13–17, 196
threshold stress, 231–235
tilt, 112, 113, 116, 134, 148, 189–191
tin–bismuth (Sn–Bi), 69, 83–86, 131
tin–copper (Sn–Cu), 69, 87–118, 126–131, 228,
    230–232, 234, 243–245

tin–lead (Sn–Pb), 81–83, 96, 161, 162, 164–167,
    183, 216–218, 220–222, 226, 230, 231
tin–silver (Sn–Ag), 69, 86–87
topography, 140–141, 143
transmission electron microscopy (TEM), 90, 93,
    94, 126–135, 142, 147, 149, 155
triple-grain junction, 48–55
triple junction, 54, 236
tungsten, 126, 127, 130, 132, 147, 188, 194, 195,
    203–209, 211–213
twinning, 141, 144, 150

underlayer, 164, 165, 228, 229, 238, 239
unwetted, 217, 218, 222

Vickers indentation, 243, 244
void, 101, 168–172, 197, 203, 218
volume per defect, 199
volume relaxation, 197, 199, 205, 208, 209

wetting, 72, 73, 217, 221, 222
whisker counting, 189
whisker density, 10, 11, 23–29, 35, 38, 49, 50,
    172, 188, 194, 215, 216
whisker diameter, 12, 203, 245
whisker length, 37, 60, 76, 78–80, 135, 139, 140,
    144, 181, 189–191, 203, 210–213, 215, 216,
    219–221, 230, 236, 238, 241–243, 245
whisker propensity, 55, 60, 75, 89–118, 160, 184,
    185
whisker radius, 10, 12, 17, 233
whisker root, 3, 4, 6, 8, 10, 12, 90, 109, 111–118,
    126, 127, 143, 150, 151
white Sn, 44

X-ray diffraction (XRD), 31, 44, 55, 58, 60, 90,
    91, 96, 102, 103, 109, 118, 147, 195, 198,
    232, 237
X-ray fluorescence (XRF), 182, 183, 192

yield stress, 27, 29, 32, 36–38, 48, 106, 240
Young's modulus, 59, 60, 102, 106, 147, 240

Z-contrast, 128, 130
zinc (Zn), 46, 90, 165, 194